RILEM State-of-the-Art Reports

RILEM STATE-OF-THE-ART REPORTS
Volume 40

RILEM, The International Union of Laboratories and Experts in Construction Materials, Systems and Structures, founded in 1947, is a non-governmental scientific association whose goal is to contribute to progress in the construction sciences, techniques and industries, essentially by means of the communication it fosters between research and practice. RILEM's focus is on construction materials and their use in building and civil engineering structures, covering all phases of the building process from manufacture to use and recycling of materials. More information on RILEM and its previous publications can be found on www.RILEM.net.

The RILEM State-of-the-Art Reports (STAR) are produced by the Technical Committees. They represent one of the most important outputs that RILEM generates – high level scientific and engineering reports that provide cutting edge knowledge in a given field. The work of the TCs is one of RILEM's key functions.

Members of a TC are experts in their field and give their time freely to share their expertise. As a result, the broader scientific community benefits greatly from RILEM's activities.

RILEM's stated objective is to disseminate this information as widely as possible to the scientific community. RILEM therefore considers the STAR reports of its TCs as of highest importance, and encourages their publication whenever possible.

The information in this and similar reports is mostly pre-normative in the sense that it provides the underlying scientific fundamentals on which standards and codes of practice are based. Without such a solid scientific basis, construction practice will be less than efficient or economical.

It is RILEM's hope that this information will be of wide use to the scientific community.

Indexed in SCOPUS, Google Scholar and SpringerLink.

Esperanza Menéndez Méndez ·
António Santos Silva
Editors

Total Alkalis in Concrete—Contribution to Alkali-silica Reaction

State-of-the-Art Report of the RILEM Technical Committee 258-AAA WP3

Springer

Editors
Esperanza Menéndez Méndez
IETcc-CSIC
Madrid, Spain

António Santos Silva
Department of Materials
National Laboratory for Civil Engineering
Lisbon, Portugal

ISSN 2213-204X ISSN 2213-2031 (electronic)
RILEM State-of-the-Art Reports
ISBN 978-3-031-56166-5 ISBN 978-3-031-56164-1 (eBook)
https://doi.org/10.1007/978-3-031-56164-1

This Springer imprint is published by the registered company Springer Nature Switzerland AG
The registered company address is: Gewerbestrasse 11, 6330 Cham, Switzerland

Paper in this product is recyclable.

While this STAR Report was in its proof-reading stages, Esperanza passed away after long illness. May she rest in peace. Esperanza has been instrumental in pursuing the preparation of this report, even long after its intended publication date. We are grateful to have known her, for her inspiration, and for her companionship. Therefore, the collective authors and editors wish to dedicate this STAR Report to her memory.

Technical Committee 258-AAA—Avoiding Alkali Aggregate Reactions in Concrete—Performance Based Concept

Chair

Børge Johannes Wigum

Deputy Chair

Jan Lindgård

TC Members

Mark G. Alexander
Mario Berra
Ingmar Borchers
Maarten A. T. M. Broekmans
Rene Brueckner
João Custódio
Vinh Dao
Mario De Rooij
Klaartje De Weerdt
Josée Duchesne
Rui Miguel Ferreira
Benoît Fournier
Sue Freitag
Ricardo Garcia-Roves Loza
Eric Giannini
Colin Giebson

Michal A. Glinicki
R. Doug Hooton
Jason H. Ideker
Patricija Kara De Maeijer
Yuichiro Kawabata
Stefan Krispel
Selmo C. Kuperman
Andreas Leemann
Lech Lewczuk
Jan Lindgård
Tung Chai Ling
Qing-Feng Liu
Renaud-Pierre Martin
Late (Esperanza Menéndez Méndez)
Birgit Meng
Urs Müller
Christoph Müller
Bård M. Pedersen
Gilles Plusquellec
Terje F. Rønning
Leandro Sanchez
António Santos Silva
Ignacio Segura
Katrin Seyfarth
Gintautas Skripkiunas
Zhenguo Shi
Ian Sims
Suvimol Sujjavanich
Late (Michael D. A. Thomas)
Børge Johannes Wigum
Late (Jonathan Wood)
Kazuo Yamada

RILEM Publications

The following list is presenting the global offer of RILEM Publications, sorted by series. Each publication is available in printed version and/or in online version.

RILEM Proceedings (PRO)

PRO 1: Durability of High Performance Concrete (ISBN: 2-912143-03-9; e-ISBN: 2-351580-12-5; e-ISBN: 2351580125); *Ed.* H. Sommer

PRO 2: Chloride Penetration into Concrete (ISBN: 2-912143-00-04; e-ISBN: 2912143454); Eds. L.-O. Nilsson and J.-P. Ollivier

PRO 3: Evaluation and Strengthening of Existing Masonry Structures (ISBN: 2-912143-02-0; e-ISBN: 2351580141); *Eds.* L. Binda and C. Modena

PRO 4: Concrete: From Material to Structure (ISBN: 2-912143-04-7; e-ISBN: 2351580206); Eds. J.-P. Bournazel and Y. Malier

PRO 5: The Role of Admixtures in High Performance Concrete (ISBN: 2-912143-05-5; e-ISBN: 2351580214); *Eds.* J. G. Cabrera and R. Rivera-Villarreal

PRO 6: High Performance Fiber Reinforced Cement Composites—HPFRCC 3 (ISBN: 2-912143-06-3; e-ISBN: 2351580222); *Eds.* H. W. Reinhardt and A. E. Naaman

PRO 7: 1st International RILEM Symposium on Self-Compacting Concrete (ISBN: 2-912143-09-8; e-ISBN: 2912143721); *Eds.* Å. Skarendahl and Ö. Petersson

PRO 8: International RILEM Symposium on Timber Engineering (ISBN: 2-912143-10-1; e-ISBN: 2351580230); *Ed.* L. Boström

PRO 9: 2nd International RILEM Symposium on Adhesion between Polymers and Concrete ISAP '99 (ISBN: 2-912143-11-X; e-ISBN: 2351580249); *Eds.* Y. Ohama and M. Puterman

PRO 10: 3rd International RILEM Symposium on Durability of Building and Construction Sealants (ISBN: 2-912143-13-6; e-ISBN: 2351580257); *Eds.* A. T. Wolf

PRO 11: 4th International RILEM Conference on Reflective Cracking in Pavements (ISBN: 2-912143-14-4; e-ISBN: 2351580265); *Eds.* A. O. Abd El Halim, D. A. Taylor and El H. H. Mohamed

PRO 12: International RILEM Workshop on Historic Mortars: Characteristics and Tests (ISBN: 2-912143-15-2; e-ISBN: 2351580273); *Eds.* P. Bartos, C. Groot and J. J. Hughes

PRO 13: 2nd International RILEM Symposium on Hydration and Setting (ISBN: 2-912143-16-0; e-ISBN: 2351580281); *Ed.* A. Nonat

PRO 14: Integrated Life-Cycle Design of Materials and Structures—ILCDES 2000 (ISBN: 951-758-408-3; e-ISBN: 235158029X); (ISSN: 0356-9403); *Ed.* S. Sarja

PRO 15: Fifth RILEM Symposium on Fibre-Reinforced Concretes (FRC)—BEFIB'2000 (ISBN: 2-912143-18-7; e-ISBN: 291214373X); *Eds.* P. Rossi and G. Chanvillard

PRO 16: Life Prediction and Management of Concrete Structures (ISBN: 2-912143-19-5; e-ISBN: 2351580303); *Ed.* D. Naus

PRO 17: Shrinkage of Concrete—Shrinkage 2000 (ISBN: 2-912143-20-9; e-ISBN: 2351580311); Eds. V. Baroghel-Bouny and P.-C. Aïtcin

PRO 18: Measurement and Interpretation of the On-Site Corrosion Rate (ISBN: 2-912143-21-7; e-ISBN: 235158032X); Eds. C. Andrade, C. Alonso, J. Fullea, J. Polimon and J. Rodriguez

PRO 19: Testing and Modelling the Chloride Ingress into Concrete (ISBN: 2-912143-22-5; e-ISBN: 2351580338); *Eds.* C. Andrade and J. Kropp

PRO 20: 1st International RILEM Workshop on Microbial Impacts on Building Materials (CD 02) (e-ISBN 978-2-35158-013-4); *Ed.* M. Ribas Silva

PRO 21: International RILEM Symposium on Connections between Steel and Concrete (ISBN: 2-912143-25-X; e-ISBN: 2351580346); *Ed.* R. Eligehausen

PRO 22: International RILEM Symposium on Joints in Timber Structures (ISBN: 2-912143-28-4; e-ISBN: 2351580354); *Eds.* S. Aicher and H.-W. Reinhardt

PRO 23: International RILEM Conference on Early Age Cracking in Cementitious Systems (ISBN: 2-912143-29-2; e-ISBN: 2351580362); Eds. K. Kovler and A. Bentur

PRO 24: 2nd International RILEM Workshop on Frost Resistance of Concrete (ISBN: 2-912143-30-6; e-ISBN: 2351580370); *Eds.* M. J. Setzer, R. Auberg and H.-J. Keck

PRO 25: International RILEM Workshop on Frost Damage in Concrete (ISBN: 2-912143-31-4; e-ISBN: 2351580389); *Eds.* D. J. Janssen, M. J. Setzer and M. B. Snyder

PRO 26: International RILEM Workshop on On-Site Control and Evaluation of Masonry Structures (ISBN: 2-912143-34-9; e-ISBN: 2351580141); *Eds.* L. Binda and R. C. de Vekey

PRO 27: International RILEM Symposium on Building Joint Sealants (CD03; e-ISBN: 235158015X); *Ed.* A. T. Wolf

PRO 28: 6th International RILEM Symposium on Performance Testing and Evaluation of Bituminous Materials—PTEBM'03 (ISBN: 2-912143-35-7; e-ISBN: 978-2-912143-77-8); *Ed.* M. N. Partl

PRO 29: 2nd International RILEM Workshop on Life Prediction and Ageing Management of Concrete Structures (ISBN: 2-912143-36-5; e-ISBN: 2912143780); *Ed.* D. J. Naus

PRO 30: 4th International RILEM Workshop on High Performance Fiber Reinforced Cement Composites—HPFRCC 4 (ISBN: 2-912143-37-3; e-ISBN: 2912143799); *Eds.* A. E. Naaman and H. W. Reinhardt

PRO 31: International RILEM Workshop on Test and Design Methods for Steel Fibre Reinforced Concrete: Background and Experiences (ISBN: 2-912143-38-1; e-ISBN: 2351580168); Eds. B. Schnütgen and L. Vandewalle

PRO 32: International Conference on Advances in Concrete and Structures 2 vol. (ISBN (set): 2-912143-41-1; e-ISBN: 2351580176); *Eds. Ying-shu Yuan, Surendra P. Shah and Heng-lin Lü*

PRO 33: 3rd International Symposium on Self-Compacting Concrete (ISBN: 2-912143-42-X; e-ISBN: 2912143713); *Eds.* Ó. Wallevik and I. Níelsson

PRO 34: International RILEM Conference on Microbial Impact on Building Materials (ISBN: 2-912143-43-8; e-ISBN: 2351580184); *Ed.* M. Ribas Silva

PRO 35: International RILEM TC 186-ISA on Internal Sulfate Attack and Delayed Ettringite Formation (ISBN: 2-912143-44-6; e-ISBN: 2912143802); *Eds.* K. Scrivener and J. Skalny

PRO 36: International RILEM Symposium on Concrete Science and Engineering—A Tribute to Arnon Bentur (ISBN: 2-912143-46-2; e-ISBN: 2912143586); Eds. K. Kovler, J. Marchand, S. Mindess and J. Weiss

PRO 37: 5th International RILEM Conference on Cracking in Pavements—Mitigation, Risk Assessment and Prevention (ISBN: 2-912143-47-0; e-ISBN: 2912143764); Eds. C. Petit, I. Al-Qadi and A. Millien

PRO 38: 3rd International RILEM Workshop on Testing and Modelling the Chloride Ingress into Concrete (ISBN: 2-912143-48-9; e-ISBN: 2912143578); *Eds.* C. Andrade and J. Kropp

PRO 39: 6th International RILEM Symposium on Fibre-Reinforced Concretes—BEFIB 2004 (ISBN: 2-912143-51-9; e-ISBN: 2912143748); *Eds.* M. Di Prisco, R. Felicetti and G. A. Plizzari

PRO 40: International RILEM Conference on the Use of Recycled Materials in Buildings and Structures (ISBN: 2-912143-52-7; e-ISBN: 2912143756); *Eds.* E. Vázquez, Ch. F. Hendriks and G. M. T. Janssen

PRO 41: RILEM International Symposium on Environment-Conscious Materials and Systems for Sustainable Development (ISBN: 2-912143-55-1; e-ISBN: 2912143640); *Eds.* N. Kashino and Y. Ohama

PRO 42: SCC'2005—China: 1st International Symposium on Design, Performance and Use of Self-Consolidating Concrete (ISBN: 2-912143-61-6; e-ISBN: 2912143624); *Eds.* Zhiwu Yu, Caijun Shi, Kamal Henri Khayat and Youjun Xie

PRO 43: International RILEM Workshop on Bonded Concrete Overlays (e-ISBN: 2-912143-83-7); *Eds.* J. L. Granju and J. Silfwerbrand

PRO 44: 2nd International RILEM Workshop on Microbial Impacts on Building Materials (CD11) (e-ISBN: 2-912143-84-5); *Ed. M. Ribas Silva*

PRO 45: 2nd International Symposium on Nanotechnology in Construction, Bilbao (ISBN: 2-912143-87-X; e-ISBN: 2912143888); *Eds. Peter J. M. Bartos, Yolanda de Miguel and Antonio Porro*

PRO 46: ConcreteLife'06—International RILEM-JCI Seminar on Concrete Durability and Service Life Planning: Curing, Crack Control, Performance in Harsh Environments (ISBN: 2-912143-89-6; e-ISBN: 291214390X); *Ed. K. Kovler*

PRO 47: International RILEM Workshop on Performance Based Evaluation and Indicators for Concrete Durability (ISBN: 978-2-912143-95-2; e-ISBN: 9782912143969); *Eds. V. Baroghel-Bouny, C. Andrade, R. Torrent and K. Scrivener*

PRO 48: 1st International RILEM Symposium on Advances in Concrete through Science and Engineering (e-ISBN: 2-912143-92-6); *Eds. J. Weiss, K. Kovler, J. Marchand, and S. Mindess*

PRO 49: International RILEM Workshop on High Performance Fiber Reinforced Cementitious Composites in Structural Applications (ISBN: 2-912143-93-4; e-ISBN: 2912143942); *Eds. G. Fischer and V.C. Li*

PRO 50: 1st International RILEM Symposium on Textile Reinforced Concrete (ISBN: 2-912143-97-7; e-ISBN: 2351580087); *Eds. Josef Hegger, Wolfgang Brameshuber and Norbert Will*

PRO 51: 2nd International Symposium on Advances in Concrete through Science and Engineering (ISBN: 2-35158-003-6; e-ISBN: 2-35158-002-8); *Eds. J. Marchand, B. Bissonnette, R. Gagné, M. Jolin and F. Paradis*

PRO 52: Volume Changes of Hardening Concrete: Testing and Mitigation (ISBN: 2-35158-004-4; e-ISBN: 2-35158-005-2); *Eds. O. M. Jensen, P. Lura and K. Kovler*

PRO 53: High Performance Fiber Reinforced Cement Composites—HPFRCC5 (ISBN: 978-2-35158-046-2; e-ISBN: 978-2-35158-089-9); *Eds. H. W. Reinhardt and A. E. Naaman*

PRO 54: 5th International RILEM Symposium on Self-Compacting Concrete (ISBN: 978-2-35158-047-9; e-ISBN: 978-2-35158-088-2); *Eds. G. De Schutter and V. Boel*

PRO 55: International RILEM Symposium Photocatalysis, Environment and Construction Materials (ISBN: 978-2-35158-056-1; e-ISBN: 978-2-35158-057-8); *Eds. P. Baglioni and L. Cassar*

PRO 56: International RILEM Workshop on Integral Service Life Modelling of Concrete Structures (ISBN 978-2-35158-058-5; e-ISBN: 978-2-35158-090-5); *Eds. R. M. Ferreira, J. Gulikers and C. Andrade*

PRO 57: RILEM Workshop on Performance of cement-based materials in aggressive aqueous environments (e-ISBN: 978-2-35158-059-2); *Ed. N. De Belie*

PRO 58: International RILEM Symposium on Concrete Modelling—CONMOD'08 (ISBN: 978-2-35158-060-8; e-ISBN: 978-2-35158-076-9); *Eds. E. Schlangen and G. De Schutter*

PRO 59: International RILEM Conference on On Site Assessment of Concrete, Masonry and Timber Structures—SACoMaTiS 2008 (ISBN set: 978-2-35158-061-5; e-ISBN: 978-2-35158-075-2); *Eds. L. Binda, M. di Prisco and R. Felicetti*

PRO 60: Seventh RILEM International Symposium on Fibre Reinforced Concrete: Design and Applications—BEFIB 2008 (ISBN: 978-2-35158-064-6; e-ISBN: 978-2-35158-086-8); *Ed. R. Gettu*

PRO 61: 1st International Conference on Microstructure Related Durability of Cementitious Composites 2 vol., (ISBN: 978-2-35158-065-3; e-ISBN: 978-2-35158-084-4); *Eds. W. Sun, K. van Breugel, C. Miao, G. Ye and H. Chen*

PRO 62: NSF/ RILEM Workshop: In-situ Evaluation of Historic Wood and Masonry Structures (e-ISBN: 978-2-35158-068-4); *Eds. B. Kasal, R. Anthony and M. Drdácký*

PRO 63: Concrete in Aggressive Aqueous Environments: Performance, Testing and Modelling, 2 vol., (ISBN: 978-2-35158-071-4; e-ISBN: 978-2-35158-082-0); *Eds. M. G. Alexander and A. Bertron*

PRO 64: Long Term Performance of Cementitious Barriers and Reinforced Concrete in Nuclear Power Plants and Waste Management—NUCPERF 2009 (ISBN: 978-2-35158-072-1; e-ISBN: 978-2-35158-087-5); *Eds. V. L'Hostis, R. Gens, C. Gallé*

PRO 65: Design Performance and Use of Self-consolidating Concrete—SCC'2009 (ISBN: 978-2-35158-073-8; e-ISBN: 978-2-35158-093-6); *Eds. C. Shi, Z. Yu, K. H. Khayat and P. Yan*

PRO 66: 2nd International RILEM Workshop on Concrete Durability and Service Life Planning—ConcreteLife'09 (ISBN: 978-2-35158-074-5; ISBN: 978-2-35158-074-5); *Ed. K. Kovler*

PRO 67: Repairs Mortars for Historic Masonry (e-ISBN: 978-2-35158-083-7); *Ed. C. Groot*

PRO 68: Proceedings of the 3rd International RILEM Symposium on 'Rheology of Cement Suspensions such as Fresh Concrete (ISBN 978-2-35158-091-2; e-ISBN: 978-2-35158-092-9); *Eds. O. H. Wallevik, S. Kubens and S. Oesterheld*

PRO 69: 3rd International PhD Student Workshop on 'Modelling the Durability of Reinforced Concrete (ISBN: 978-2-35158-095-0); *Eds. R. M. Ferreira, J. Gulikers and C. Andrade*

PRO 70: 2nd International Conference on 'Service Life Design for Infrastructure' (ISBN set: 978-2-35158-096-7, e-ISBN: 978-2-35158-097-4); *Ed. K. van Breugel, G. Ye and Y. Yuan*

PRO 71: Advances in Civil Engineering Materials—The 50-year Teaching Anniversary of Prof. Sun Wei' (ISBN: 978-2-35158-098-1; e-ISBN: 978-2-35158-099-8); *Eds. C. Miao, G. Ye, and H. Chen*

PRO 72: First International Conference on 'Advances in Chemically-Activated Materials—CAM'2010' (2010), 264 pp, ISBN: 978-2-35158-101-8; e-ISBN: 978-2-35158-115-5, *Eds. Caijun Shi and Xiaodong Shen*

PRO 73: 2nd International Conference on 'Waste Engineering and Management—ICWEM 2010' (2010), 894 pp, ISBN: 978-2-35158-102-5; e-ISBN: 978-2-35158-103-2, *Eds. J. Zh. Xiao, Y. Zhang, M. S. Cheung and R. Chu*

PRO 74: International RILEM Conference on 'Use of Superabsorsorbent Polymers and Other New Addditives in Concrete' (2010) 374 pp., ISBN: 978-2-35158-104-9; e-ISBN: 978-2-35158-105-6; *Eds. O.M. Jensen, M.T. Hasholt, and S. Laustsen*

PRO 75: International Conference on 'Material Science—2nd ICTRC—Textile Reinforced Concrete—Theme 1' (2010) 436 pp., ISBN: 978-2-35158-106-3; e-ISBN: 978-2-35158-107-0; *Ed. W. Brameshuber*

PRO 76: International Conference on 'Material Science—HetMat—Modelling of Heterogeneous Materials—Theme 2' (2010) 255 pp., ISBN: 978-2-35158-108-7; e-ISBN: 978-2-35158-109-4; *Ed. W. Brameshuber*

PRO 77: International Conference on 'Material Science—AdIPoC—Additions Improving Properties of Concrete—Theme 3' (2010) 459 pp., ISBN: 978-2-35158-110-0; e-ISBN: 978-2-35158-111-7; *Ed. W. Brameshuber*

PRO 78: 2nd Historic Mortars Conference and RILEM TC 203-RHM Final Workshop—HMC2010 (2010) 1416 pp., e-ISBN: 978-2-35158-112-4; *Eds J. Válek, C. Groot, and J. J. Hughes*

PRO 79: International RILEM Conference on Advances in Construction Materials Through Science and Engineering (2011) 213 pp., ISBN: 978-2-35158-116-2, e-ISBN: 978-2-35158-117-9; *Eds Christopher Leung and K.T. Wan*

PRO 80: 2nd International RILEM Conference on Concrete Spalling due to Fire Exposure (2011) 453 pp., ISBN: 978-2-35158-118-6, e-ISBN: 978-2-35158-119-3; *Eds E.A.B. Koenders and F. Dehn*

PRO 81: 2nd International RILEM Conference on Strain Hardening Cementitious Composites (SHCC2-Rio) (2011) 451 pp., ISBN: 978-2-35158-120-9, e-ISBN: 978-2-35158-121-6; *Eds R.D. Toledo Filho, F.A. Silva, E.A.B. Koenders and E.M.R. Fairbairn*

PRO 82: 2nd International RILEM Conference on Progress of Recycling in the Built Environment (2011) 507 pp., e-ISBN: 978-2-35158-122-3; *Eds V.M. John, E. Vazquez, S.C. Angulo and C. Ulsen*

PRO 83: 2nd International Conference on Microstructural-related Durability of Cementitious Composites (2012) 250 pp., ISBN: 978-2-35158-129-2; e-ISBN: 978-2-35158-123-0; *Eds G. Ye, K. van Breugel, W. Sun and C. Miao*

PRO 84: CONSEC13 - Seventh International Conference on Concrete under Severe Conditions—Environment and Loading (2013) 1930 pp., ISBN: 978-2-35158-124-7; e-ISBN: 978-2- 35158-134-6; *Eds Z.J. Li, W. Sun, C.W. Miao, K. Sakai, O.E. Gjorv & N. Banthia*

PRO 85: RILEM-JCI International Workshop on Crack Control of Mass Concrete and Related issues concerning Early-Age of Concrete Structures—ConCrack 3—Control of Cracking in Concrete Structures 3 (2012) 237 pp., ISBN: 978-2-35158-125-4; e-ISBN: 978-2-35158-126-1; *Eds F. Toutlemonde and J.-M. Torrenti*

PRO 86: International Symposium on Life Cycle Assessment and Construction (2012) 414 pp., ISBN: 978-2-35158-127-8, e-ISBN: 978-2-35158-128-5; *Eds A. Ventura and C. de la Roche*

PRO 87: UHPFRC 2013—RILEM-fib-AFGC International Symposium on Ultra-High Performance Fibre-Reinforced Concrete (2013), ISBN: 978-2-35158-130-8, e-ISBN: 978-2-35158-131-5; *Eds F. Toutlemonde*

PRO 88: 8th RILEM International Symposium on Fibre Reinforced Concrete (2012) 344 pp., ISBN: 978-2-35158-132-2, e-ISBN: 978-2-35158-133-9; *Eds Joaquim A.O. Barros*

PRO 89: RILEM International workshop on performance-based specification and control of concrete durability (2014) 678 pp, ISBN: 978-2-35158-135-3, e-ISBN: 978-2-35158-136-0; *Eds. D. Bjegović, H. Beushausen and M. Serdar*

PRO 90: 7th RILEM International Conference on Self-Compacting Concrete and of the 1st RILEM International Conference on Rheology and Processing of Construction Materials (2013) 396 pp, ISBN: 978-2-35158-137-7, e-ISBN: 978-2-35158-138-4; *Eds. Nicolas Roussel and Hela Bessaies-Bey*

PRO 91: CONMOD 2014—RILEM International Symposium on Concrete Modelling (2014), ISBN: 978-2-35158-139-1; e-ISBN: 978-2-35158-140-7; *Eds. Kefei Li, Peiyu Yan and Rongwei Yang*

PRO 92: CAM 2014—2nd International Conference on advances in chemically-activated materials (2014) 392 pp., ISBN: 978-2-35158-141-4; e-ISBN: 978-2-35158-142-1; *Eds. Caijun Shi and Xiadong Shen*

PRO 93: SCC 2014—3rd International Symposium on Design, Performance and Use of Self-Consolidating Concrete (2014) 438 pp., ISBN: 978-2-35158-143-8; e-ISBN: 978-2-35158-144-5; *Eds. Caijun Shi, Zhihua Ou, Kamal H. Khayat*

PRO 94 (online version): HPFRCC-7—7th RILEM conference on High performance fiber reinforced cement composites (2015), e-ISBN: 978-2-35158-146-9; *Eds. H.W. Reinhardt, G.J. Parra-Montesinos, H. Garrecht*

PRO 95: International RILEM Conference on Application of superabsorbent polymers and other new admixtures in concrete construction (2014), ISBN: 978-2-35158-147-6; e-ISBN: 978-2-35158-148-3; *Eds. Viktor Mechtcherine, Christof Schroefl*

PRO 96 (online version): XIII DBMC: XIII International Conference on Durability of Building Materials and Components(2015), e-ISBN: 978-2-35158-149-0; *Eds. M. Quattrone, V.M. John*

PRO 97: SHCC3—3rd International RILEM Conference on Strain Hardening Cementitious Composites (2014), ISBN: 978-2-35158-150-6; e-ISBN: 978-2-35158-151-3; *Eds. E. Schlangen, M.G. Sierra Beltran, M. Lukovic, G. Ye*

PRO 98: FERRO-11—11th International Symposium on Ferrocement and 3rd ICTRC—International Conference on Textile Reinforced Concrete (2015), ISBN: 978-2-35158-152-0; e-ISBN: 978-2-35158-153-7; *Ed. W. Brameshuber*

PRO 99 (online version): ICBBM 2015—1st International Conference on Bio-Based Building Materials (2015), e-ISBN: 978-2-35158-154-4; *Eds. S. Amziane, M. Sonebi*

PRO 100: SCC16—RILEM Self-Consolidating Concrete Conference (2016), ISBN: 978-2-35158-156-8; e-ISBN: 978-2-35158-157-5; *Ed. Kamal H. Kayat*

PRO 101 (online version): III Progress of Recycling in the Built Environment (2015), e-ISBN: 978-2-35158-158-2; *Eds I. Martins, C. Ulsen and S. C. Angulo*

PRO 102 (online version): RILEM Conference on Microorganisms-Cementitious Materials Interactions (2016), e-ISBN: 978-2-35158-160-5; *Eds. Alexandra Bertron, Henk Jonkers, Virginie Wiktor*

PRO 103 (online version): ACESC'16—Advances in Civil Engineering and Sustainable Construction (2016), e-ISBN: 978-2-35158-161-2; *Eds. T.Ch. Madhavi, G. Prabhakar, Santhosh Ram and P.M. Rameshwaran*

PRO 104 (online version): SSCS'2015—Numerical Modeling—Strategies for Sustainable Concrete Structures (2015), e-ISBN: 978-2-35158-162-9

PRO 105: 1st International Conference on UHPC Materials and Structures (2016), ISBN: 978-2-35158-164-3, e-ISBN: 978-2-35158-165-0

PRO 106: AFGC-ACI-fib-RILEM International Conference on Ultra-High-Performance Fibre-Reinforced Concrete—UHPFRC 2017 (2017), ISBN: 978-2-35158-166-7, e-ISBN: 978-2-35158-167-4; *Eds. François Toutlemonde & Jacques Resplendino*

PRO 107 (online version): XIV DBMC—14th International Conference on Durability of Building Materials and Components (2017), e-ISBN: 978-2-35158-159-9; *Eds. Geert De Schutter, Nele De Belie, Arnold Janssens, Nathan Van Den Bossche*

PRO 108: MSSCE 2016—Innovation of Teaching in Materials and Structures (2016), ISBN: 978-2-35158-178-0, e-ISBN: 978-2-35158-179-7; *Ed. Per Goltermann*

PRO 109 (2 volumes): MSSCE 2016—Service Life of Cement-Based Materials and Structures (2016), ISBN Vol. 1: 978-2-35158-170-4, Vol. 2: 978-2-35158-171-4, Set Vol. 1&2: 978-2-35158-172-8, e-ISBN : 978-2-35158-173-5; *Eds. Miguel Azenha, Ivan Gabrijel, Dirk Schlicke, Terje Kanstad and Ole Mejlhede Jensen*

PRO 110: MSSCE 2016—Historical Masonry (2016), ISBN: 978-2-35158-178-0, e-ISBN: 978-2-35158-179-7; *Eds. Inge Rörig-Dalgaard and Ioannis Ioannou*

PRO 111: MSSCE 2016—Electrochemistry in Civil Engineering (2016), ISBN: 978-2-35158-176-6, e-ISBN: 978-2-35158-177-3; *Ed. Lisbeth M. Ottosen*

PRO 112: MSSCE 2016—Moisture in Materials and Structures (2016), ISBN: 978-2-35158-178-0, e-ISBN: 978-2-35158-179-7; *Eds. Kurt Kielsgaard Hansen, Carsten Rode and Lars-Olof Nilsson*

PRO 113: MSSCE 2016—Concrete with Supplementary Cementitious Materials (2016), ISBN: 978-2-35158-178-0, e-ISBN: 978-2-35158-179-7; *Eds. Ole Mejlhede Jensen, Konstantin Kovler and Nele De Belie*

PRO 114: MSSCE 2016—Frost Action in Concrete (2016), ISBN: 978-2-35158-182-7, e-ISBN: 978-2-35158-183-4; *Eds. Marianne Tange Hasholt, Katja Fridh and R. Doug Hooton*

PRO 115: MSSCE 2016—Fresh Concrete (2016), ISBN: 978-2-35158-184-1, e-ISBN: 978-2-35158-185-8; *Eds. Lars N. Thrane, Claus Pade, Oldrich Svec and Nicolas Roussel*

PRO 116: BEFIB 2016—9th RILEM International Symposium on Fiber Reinforced Concrete (2016), ISBN: 978-2-35158-187-2, e-ISBN: 978-2-35158-186-5; *Eds. N. Banthia, M. di Prisco and S. Soleimani-Dashtaki*

PRO 117: 3rd International RILEM Conference on Microstructure Related Durability of Cementitious Composites (2016), ISBN: 978-2-35158-188-9, e-ISBN: 978-2-35158-189-6; *Eds. Changwen Miao, Wei Sun, Jiaping Liu, Huisu Chen, Guang Ye and Klaas van Breugel*

PRO 118 (4 volumes): International Conference on Advances in Construction Materials and Systems (2017), ISBN Set: 978-2-35158-190-2, Vol. 1: 978-2-35158-193-3, Vol. 2: 978-2-35158-194-0, Vol. 3: ISBN:978-2-35158-195-7, Vol. 4: ISBN:978-2-35158-196-4, e-ISBN: 978-2-35158-191-9; *Eds. Manu Santhanam, Ravindra Gettu, Radhakrishna G. Pillai and Sunitha K. Nayar*

PRO 119 (online version): ICBBM 2017—Second International RILEM Conference on Bio-based Building Materials, (2017), e-ISBN: 978-2-35158-192-6; *Eds. Sofiane Amziane, Mohammed Sonebi*

PRO 120 (2 volumes): EAC-02—2nd International RILEM/COST Conference on Early Age Cracking and Serviceability in Cement-based Materials and Structures, (2017), Vol. 1: 978-2-35158-199-5, Vol. 2: 978-2-35158-200-8, Set: 978-2-35158-197-1, e-ISBN: 978-2-35158-198-8; *Eds. Stéphanie Staquet and Dimitrios Aggelis*

PRO 121 (2 volumes): SynerCrete18: Interdisciplinary Approaches for Cement-based Materials and Structural Concrete: Synergizing Expertise and Bridging Scales of Space and Time, (2018), Set: 978-2-35158-202-2, Vol.1: 978-2-35158-211-4, Vol.2: 978-2-35158-212-1, e-ISBN: 978-2-35158-203-9; *Eds. Miguel Azenha, Dirk Schlicke, Farid Benboudjema, Agnieszka Knoppik*

PRO 122: SCC'2018 China—Fourth International Symposium on Design, Performance and Use of Self-Consolidating Concrete, (2018), ISBN: 978-2-35158-204-6, e-ISBN: 978-2-35158-205-3; *Eds. C. Shi, Z. Zhang, K. H. Khayat*

PRO 123: Final Conference of RILEM TC 253-MCI: Microorganisms-Cementitious Materials Interactions (2018), Set: 978-2-35158-207-7, Vol.1: 978-2-35158-209-1, Vol.2: 978-2-35158-210-7, e-ISBN: 978-2-35158-206-0; *Ed. Alexandra Bertron*

PRO 124 (online version): Fourth International Conference Progress of Recycling in the Built Environment (2018), e-ISBN: 978-2-35158-208-4; *Eds. Isabel M. Martins, Carina Ulsen, Yury Villagran*

PRO 125 (online version): SLD4—4th International Conference on Service Life Design for Infrastructures (2018), e-ISBN: 978-2-35158-213-8; *Eds. Guang Ye, Yong Yuan, Claudia Romero Rodriguez, Hongzhi Zhang, Branko Savija*

PRO 126: Workshop on Concrete Modelling and Material Behaviour in honor of Professor Klaas van Breugel (2018), ISBN: 978-2-35158-214-5, e-ISBN: 978-2-35158-215-2; *Ed. Guang Ye*

PRO 127 (online version): CONMOD2018—Symposium on Concrete Modelling (2018), e-ISBN: 978-2-35158-216-9; *Eds. Erik Schlangen, Geert de Schutter, Branko Savija, Hongzhi Zhang, Claudia Romero Rodriguez*

PRO 128: SMSS2019—International Conference on Sustainable Materials, Systems and Structures (2019), ISBN: 978-2-35158-217-6, e-ISBN: 978-2-35158-218-3

PRO 129: 2nd International Conference on UHPC Materials and Structures (UHPC2018-China), ISBN: 978-2-35158-219-0, e-ISBN: 978-2-35158-220-6;

PRO 130: 5th Historic Mortars Conference (2019), ISBN: 978-2-35158-221-3, e-ISBN: 978-2-35158-222-0; *Eds. José Ignacio Álvarez, José María Fernández, Íñigo Navarro, Adrián Durán, Rafael Sirera*

PRO 131 (online version): 3rd International Conference on Bio-Based Building Materials (ICBBM2019), e-ISBN: 978-2-35158-229-9; *Eds. Mohammed Sonebi, Sofiane Amziane, Jonathan Page*

PRO 132: IRWRMC'18—International RILEM Workshop on Rheological Measurements of Cement-based Materials (2018), ISBN: 978-2-35158-230-5, e-ISBN: 978-2-35158-231-2; *Eds. Chafika Djelal, Yannick Vanhove*

PRO 133 (online version): CO2STO2019—International Workshop CO2 Storage in Concrete (2019), e-ISBN: 978-2-35158-232-9; *Eds. Assia Djerbi, Othman Omikrine-Metalssi, Teddy Fen-Chong*

PRO 134: 3rd ACF/HNU International Conference on UHPC Materials and Structures—UHPC'2020, ISBN: 978-2-35158-233-6, e-ISBN: 978-2-35158-234-3; *Eds. Caijun Shi & Jiaping Liu*

PRO 135: Fourth International Conference on Chemically Activated Materials (CAM2021), ISBN: 978-2-35158-235-0, e-ISBN: 978-2-35158-236-7; *Eds. Caijun Shi & Xiang Hu*

PRO 136: Lime Based Materials for Repairing Historic Structures (2022), ISBN: 978-2-35158-240-4, e-ISBN: 978-2-35158-237-4; *Eds. Ioanna Papayianni & Jan Válek*

PRO 137 (online version): V International Conference Progress of Recycling in the Built Environment (2023), e-ISBN: 978-2-35158-238-1; *Eds. Ines Döring, Yury Villagrán Zaccardi, Anette Müller*

RILEM Reports (REP)

Report 19: Considerations for Use in Managing the Aging of Nuclear Power Plant Concrete Structures (ISBN: 2-912143-07-1); *Ed. D. J. Naus*

Report 20: Engineering and Transport Properties of the Interfacial Transition Zone in Cementitious Composites (ISBN: 2-912143-08-X); Eds. M. G. Alexander, G. Arliguie, G. Ballivy, A. Bentur and J. Marchand

Report 21: Durability of Building Sealants (ISBN: 2-912143-12-8); *Ed.* A. T. Wolf

Report 22: Sustainable Raw Materials—Construction and Demolition Waste (ISBN: 2-912143-17-9); *Eds.* C. F. Hendriks and H. S. Pietersen

Report 23: Self-Compacting Concrete state-of-the-art report (ISBN: 2-912143-23-3); *Eds.* Å. Skarendahl and Ö. Petersson

Report 24: Workability and Rheology of Fresh Concrete: Compendium of Tests (ISBN: 2-912143-32-2); *Eds.* P. J. M. Bartos, M. Sonebi and A. K. Tamimi

Report 25: Early Age Cracking in Cementitious Systems (ISBN: 2-912143-33-0); Ed. A. Bentur

Report 26: Towards Sustainable Roofing (Joint Committee CIB/RILEM) (CD 07) (e-ISBN 978-2-912143-65-5); *Eds. Thomas W. Hutchinson and Keith Roberts*

Report 27: Condition Assessment of Roofs (Joint Committee CIB/RILEM) (CD 08) (e-ISBN 978-2-912143-66-2); *Ed. CIB W 83/RILEM TC166-RMS*

Report 28: Final report of RILEM TC 167-COM 'Characterisation of Old Mortars with Respect to Their Repair (ISBN: 978-2-912143-56-3); *Eds.* C. Groot, G. Ashall and J. Hughes

Report 29: Pavement Performance Prediction and Evaluation (PPPE): Interlaboratory Tests (e-ISBN: 2-912143-68-3); *Eds.* M. Partl and H. Piber

Report 30: Final Report of RILEM TC 198-URM 'Use of Recycled Materials' (ISBN: 2-912143-82-9; e-ISBN: 2-912143-69-1); *Eds.* Ch. F. Hendriks, G. M. T. Janssen and E. Vázquez

Report 31: Final Report of RILEM TC 185-ATC 'Advanced testing of cement-based materials during setting and hardening' (ISBN: 2-912143-81-0; e-ISBN: 2-912143-70-5); *Eds.* H. W. Reinhardt and C. U. Grosse

Report 32: Probabilistic Assessment of Existing Structures. A JCSS publication (ISBN 2-912143-24-1); Ed. D. Diamantidis

Report 33: State-of-the-Art Report of RILEM Technical Committee TC 184-IFE 'Industrial Floors' (ISBN 2-35158-006-0); Ed. P. Seidler

Report 34: Report of RILEM Technical Committee TC 147-FMB 'Fracture mechanics applications to anchorage and bond' Tension of Reinforced Concrete Prisms—Round Robin Analysis and Tests on Bond (e-ISBN 2-912143-91-8); Eds. L. Elfgren and K. Noghabai

Report 35: Final Report of RILEM Technical Committee TC 188-CSC 'Casting of Self Compacting Concrete' (ISBN 2-35158-001-X; e-ISBN: 2-912143-98-5); Eds. Å. Skarendahl and P. Billberg

Report 36: State-of-the-Art Report of RILEM Technical Committee TC 201-TRC 'Textile Reinforced Concrete' (ISBN 2-912143-99-3); Ed. W. Brameshuber

Report 37: State-of-the-Art Report of RILEM Technical Committee TC 192-ECM 'Environment-conscious construction materials and systems' (ISBN: 978-2-35158-053-0); Eds. N. Kashino, D. Van Gemert and K. Imamoto

Report 38: State-of-the-Art Report of RILEM Technical Committee TC 205-DSC 'Durability of Self-Compacting Concrete' (ISBN: 978-2-35158-048-6); Eds. G. De Schutter and K. Audenaert

Report 39: Final Report of RILEM Technical Committee TC 187-SOC 'Experimental determination of the stress-crack opening curve for concrete in tension' (ISBN 978-2-35158-049-3); Ed. J. Planas

Report 40: State-of-the-Art Report of RILEM Technical Committee TC 189-NEC 'Non-Destructive Evaluation of the Penetrability and Thickness of the Concrete Cover' (ISBN 978-2-35158-054-7); Eds. R. Torrent and L. Fernández Luco

Report 41: State-of-the-Art Report of RILEM Technical Committee TC 196-ICC 'Internal Curing of Concrete' (ISBN 978-2-35158-009-7); Eds. K. Kovler and O. M. Jensen

Report 42: 'Acoustic Emission and Related Non-destructive Evaluation Techniques for Crack Detection and Damage Evaluation in Concrete'—Final Report of RILEM Technical Committee 212-ACD (e-ISBN: 978-2-35158-100-1); Ed. M. Ohtsu

Report 45: Repair Mortars for Historic Masonry—State-of-the-Art Report of RILEM Technical Committee TC 203-RHM (e-ISBN: 978-2-35158-163-6); Eds. Paul Maurenbrecher and Caspar Groot

Report 46: Surface delamination of concrete industrial floors and other durability related aspects guide—Report of RILEM Technical Committee TC 268-SIF (e-ISBN: 978-2-35158-201-5); Ed. Valerie Pollet

Contents

Chapter 1
Introduction

Esperanza Menéndez Méndez and António Santos Silva

The alkali-aggregate reaction (AAR) is an expansive chemical reaction that occurs in cementitious materials when certain minerals present in the aggregates react in the presence of water with the hydroxyl ions (OH^-) and the alkalis[1] present in the pore solution. In AAR there are generally included two reaction types: alkali-silica reaction (ASR) and alkali-carbonate reaction (ACR), being the ACR extremely rare [1]. Due to the ASR prevalence, this type of AAR will be the focus in this document.

Thomas Stanton was the first to publish a paper on ASR, in November, 1940 in the proceedings of the American Society of Civil Engineers implicating the alkalis in cement and hydrous silica in aggregates as participants in a chemical reaction that appeared to have caused expansion and cracking of a number of concrete structures in Monterey County (Fig. 1.1). Shortly after publication of this paper ASR was diagnosed as being the cause of expansion and cracking in several other structures, including the Parker Dam and Stewart Mountain Dan in Arizona.

The speed of the ASR is very dependent on the type and content of reactive mineral present (usually silica-containing) in the aggregates, as well as the availability of alkalis and water, which explains why in certain cases the symptoms in the structures only manifest after 30 years or more after its entry into service. Therefore, many new cases of concrete structures of a certain age are currently being diagnosed.

---

[1] In this document, the term alkalis refers to the alkali metal sodium (Na) and potassium (K).

Esperanza Menéndez Méndez—Deceased 21 November 2024.

E. Menéndez Méndez (Deceased)
Institute Eduardo Torroja of Construction Science (IETcc-CSIC), C/Serrano Galvache, 4, 28033 Madrid, Spain
e-mail: emm@ietcc.csic.es

A. Santos Silva (✉)
National Laboratory for Civil Engineering (LNEC), Av. do Brasil 101, 1700-066 Lisboa, Portugal
e-mail: ssilva@lnec.pt

E. Menéndez Méndez and A. Santos Silva (eds.), *Total Alkalis in Concrete—Contribution to Alkali-silica Reaction*, RILEM State-of-the-Art Reports 40, https://doi.org/10.1007/978-3-031-56164-1_1

For ASR to occur, it is necessary that there are simultaneously potentially reactive siliceous minerals in the aggregates, enough free alkalis and water (Fig. 1.2). Alkalis are mainly contributed from the cement, and in minor amount from the aggregates, deicing salts, additions, and use of saline waters.

The main reaction product of ASR is an alkaline silicate gel that will swell with the imbibition of water. The gels flow into the cementitious paste filling voids and pores (Fig. 1.3) and initially densifying the paste. Crack formation occurs when the

Fig. 1.1 First image of Thomas Stanton with a structure affected by alkali-silica reaction [5]

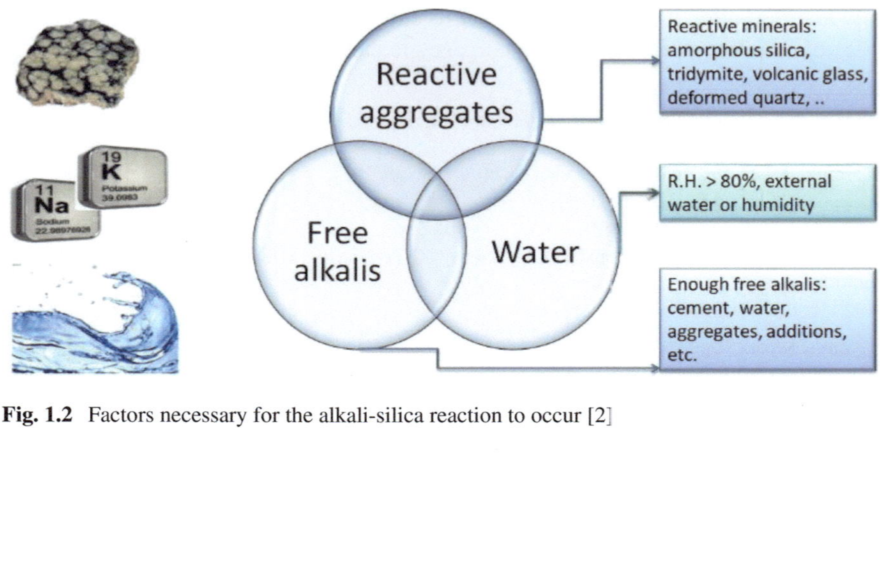

Fig. 1.2 Factors necessary for the alkali-silica reaction to occur [2]

Fig. 1.3 Microstructural aspect of silica gel and other products of ASR by stereomicroscopy [2]

tensile strength of the paste or aggregate is exceeded. However, the ASR mechanism is complex and evolves over time, with the possibility of ion exchanges with calcium or other ions that modify the chemical composition of the gels (Fig. 1.4). This modification in the composition can cause them to stabilize, but also that the released alkalis are recycled in the paste, causing the reaction to continue.

Expansive phenomena extend over time depending on many factors, such as the type of aggregates, the type and amount of cement, free water in the concrete, environmental conditions (Fig. 1.5), as well as the distribution of mechanical loads in different areas of a same structure.

The expansion due to the ASR behaves like other expansive reactions (in the form of “S”), with an initiation period, a propagation period and an attenuation period (Fig. 1.6). The initiation period can be more or less long depending on the factors mentioned above (characteristics of the concrete components, environmental factors and mechanical loads). The propagation period is in which the reaction develops more rapidly, and more cracking is manifested. And, in the attenuation period is when the reaction slows down significantly, due to one of the reaction components, reactive silica, alkalis or water, is consumed.

Generally, the attenuation period is not reached due to one of the following reasons:

- The concrete is too cracked and is replaced to ensure the safety of the element or structure;
- The concrete element is easy to replace, and it is done to avoid any damage;
- Monitoring indicates the need for intervention previously to the attenuation period.

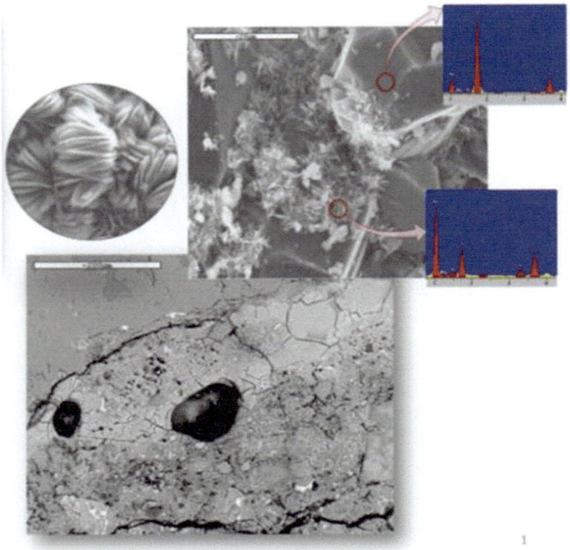

Fig. 1.4 Microstructural aspect of an ASR microcrystalline product. Upper part, SEM-EDS of microcrystals inside of a siliceous aggregate. Lower part, BSE image of gels in a paste-aggregate interface [2]

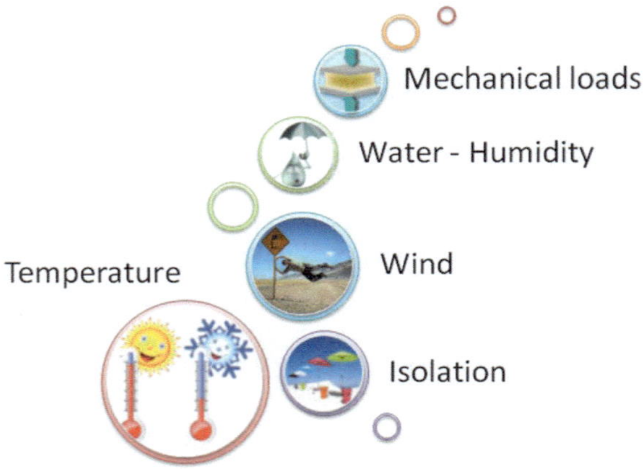

Fig. 1.5 Environmental aspects that influence the alkali-aggregate reaction development [4]

All these situations have in common the need to maintain structural safety and avoid risks to people or property.

The attenuation or stop of the expansion can be considered only when the monitoring indicates the attenuation or stopping of the phenomenon. Although it would have to be corroborated with a detailed inspection of the structure. The attenuation can occur, as indicated above, either by the consumption of one of the reagents or by an intervention that can stop the process. These interventions are usually very expensive and are only justified in unique and particularly relevant concrete structures, such as dams, nuclear power plants, bridges, etc.

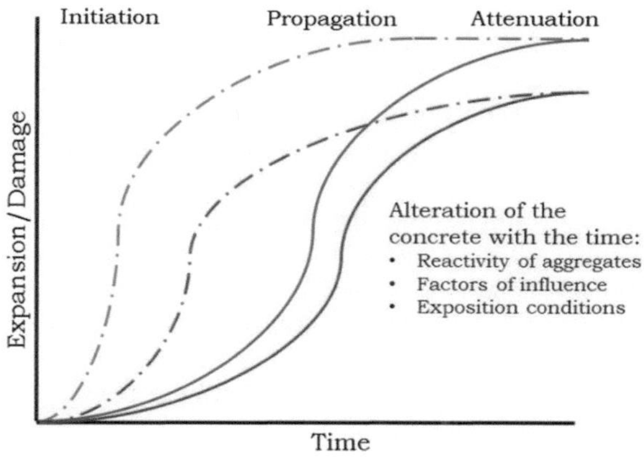

Fig. 1.6 Curve in "S" for concretes with different reaction rate [3]

As mentioned, one of the controlling components of the ASR is the alkalis available in the concrete. Traditionally, only those of the cement have been considered as free alkalis. But, given the low percentage of alkalis necessary to initiate the expansive reaction, all possible contributions that may occur in the concrete must be considered. Besides, over the past years several cases of concrete structures have been diagnosed to be affected by ASR, where the increase in the alkalis content in the interstitial solution of the cement paste has been judged to be an important contributor.

This state-of-the-art report (STAR) includes various aspects related to the total contribution of effective alkalis to concrete. This STAR was first prepared within the RILEM Technical Committee 258-AAA: Avoiding alkali aggregate reactions in concrete - Performance based concept, chaired by Professor Børge Johannes Wigum, and with the contribution of several colleagues of the TC 258-AAA.

In Chap. 2 the possible sources of internal and external alkalis are described. The internal sources include the Portland cement clinker, mixing water, chemical admixtures, aggregates, supplementary cementitious materials, fibres and the alkalis recycled due to ASR development. As for external sources, alkaline waters, sea water, fertilizers, industrial wastes, de-icing and anti-icing salts, as well as electrochemical treatments are described.

Chapter 3 deals with the analytical methods that can be used to detect alkalis in different components of concrete and in the concrete itself. In this case, it is specifically considered which procedures are used to determine total alkalis or those that can measure the effective alkalis of the concrete. The aspects related to sample size, its representativity and preparation, that are of the utmost importance for the success of the analysis, are also detailed in this chapter.

The penultimate chapter, Chap. 4, includes the international standards and recommendations for controlling the concrete alkali loadings aiming the prevention of the ASR in concrete. These include standardized methods to quantify the total amount of alkalis of the hardened concrete and its constituents, and standards to determine the releasable alkalis from the aggregates and from supplementary cementitious materials.

Finally, in annex, field cases of ASR in pavements and in concrete dams are collected, with emphasis on external and internal alkali loadings.

The different chapters presented in this STAR cover the various aspects related to the influence of alkali metals (sodium and potassium) on the ASR in concrete, but also on the analytical methods for its evaluation and on the standards and methods for its control. We feel that this book makes an important contribution to the role of alkalis in the complex mechanism of ASR. However, some aspects deserve further investigation, namely regarding the relationship between the results obtained from the analysis of concrete constituents and those carried out on the concrete itself, and between the concrete produced in the laboratory and the field concrete. The factors that lead to the release of alkalis by the aggregates are also a topic that deserves further investigation.

References

1. Godart, B., Rooij, M., Wood, J.G.M.: Guide to diagnosis and appraisal of AAR damage to concrete in structures. In: Part 1 Diagnosis (AAR 6.1), 12 RILEM State-of-the-Art Reports, Ed., 89 pp. Springer (2013). ISBN-10:9400796269, ISBN-13:978-9400796263
2. Menéndez, E.: Análisis del hormigón en estructuras afectadas por reacción árido-álcali, ataque por sulfatos y ciclos de hielo-deshielo (Analysis of concrete in structures affected by alkali-aggregate reaction, sulfate attack and freeze-thaw cycles). Ed. IECA, 233 pp. (2010). ISBN 978-84-89702-24-0
3. Menéndez, E.: Predicción del comportamiento expansivo de presas afectadas por reacción árido-álcali durante su vida útil debido al grado de reacción y parámetros microestructurales (Prediction of the expansive behaviour in dams affected by alkali-aggregate reaction during the service life due to the degree of reaction and microsctructural parameters). X Jornadas Españolas de Presas, Spancold, Sevilla, Spain (2015)
4. Menéndez, E.: Critical parameters that lead to the evolution of ASR. In: Workshop International Commission on Large Dams (ICOLD Annual meeting), Johannesburg, South Africa (2016)
5. Stanton, T.E.: Expansion of concrete through reaction between cement and aggregate. Proc. Am. Soc. Civil Eng. (ASCE) **66**, 1791–1811 (1940)

Chapter 2
Sources of Alkalis

Josée Duchesne and Isabel Fernandes

Abstract Alkalis in concrete comes from a variety of sources, both internal and external. All concrete constituents can contribute alkalis, but Portland cement is the major source of alkalis to concrete. These alkalis come from the raw materials used to make clinker and from fuels. Only a portion of the alkalis in cement are readily soluble and are highly dependent on the constituents of the clinker. Sulfate alkalis are considered readily soluble. Alkalis can come from concrete constituents such as mixing water, admixtures, aggregates, and SCMs, fibers, and can even be recycled through the products of the alkali-silica reaction. Alkalis can also come from external sources such as alkaline or sea water, fertilizers, industrial residues and de-icing products. Finally, electrochemical treatments affect the distribution of alkalis within the material. There are many sources of alkalis, and the alkalis of each component must be considered.

Keywords Alkalis · Internal sources · External sources · Alkali recycling · ASR · Electrochemical treatments

This chapter was written with the cooperation of Rene Brueckner, Cédric Drolet, Ricardo García-Rovés, Colin Giebson, Esperanza Menéndez Méndez, Leandro Sanchez, Diego de Sousa, Klaartje de Weerdt and António Santos Silva.

J. Duchesne (✉)
Université Laval, Québec, Canada
e-mail: Josee.duchesne@ggl.ulaval.ca

I. Fernandes
University of Lisbon, FCUL & IDL, Lisbon, Portugal
e-mail: mifernandes@fc.ul.pt

E. Menéndez Méndez and A. Santos Silva (eds.), *Total Alkalis in Concrete—Contribution to Alkali-silica Reaction*, RILEM State-of-the-Art Reports 40, https://doi.org/10.1007/978-3-031-56164-1_2

2.1 Internal Sources

Portland cement is the most important source of alkalis to the concrete. The alkali content of the Portland cement varies considerably and depends on the composition of the raw materials used in its manufacture, e.g., clays, shales, limestone, etc. All these materials contain different proportions of alkalis (K^+ and Na^+).

2.1.1 Clinker

The alkali contents (Na_2O_{eq}, or ($Na_2O + 0.658\ K_2O$)) usually found in ordinary Portland cements ranges from 0.3 to 1.3 wt% [88]. These alkalis originate from the raw materials used for clinker manufacture and from fuel. Table 2.1 presents typical alkali content found in raw materials.

During the heating of the raw material to form the clinker, the alkalis first bind to sulfides derived from undesirable minerals constituting shales and clays to form arcanite (potassium sulfate (K_2SO_4)), aphthitalite (sodium–potassium sulfate ($(K, Na)_2SO_4$)) and Ca-langbeinite (calcium-potassium sulfate ($K_2Ca_2(SO_4)_3$)). If the amount of sulfate in the raw material is not sufficient to bind all the alkalis, the latter are incorporated into the composition of C_3A, C_2S and, to a lesser extent, C_3S and C_4AF [88]. Woermann et al. [182] reported that up to 1.4 wt% of either Na_2O or K_2O could be incorporated into C_3S at 1500 °C. These values are high compared to typical compositions of phases in Portland cement clinkers reported by Jawed and Skalny [88] or Taylor [164] (Table 2.2). Taylor [164] also reported an upper limit of 5.7 wt% Na_2O substitution in C_3A in the absence of another substituent.

According to Jawed and Skalny [88], about 50 wt% of alkalis present in the raw materials are volatilized in the kiln between 800 and 1000 °C and are partly condensed later. A general trend is observed where retention of alkalis is higher as the energy efficiency of the kiln increases. Alkalis can also be found in other compounds in cement manufacturing such as in cement kiln dust (CKD). A part of the CKD produced is recycled back into the cement production process, but some could not be returned to the kiln due to its high alkaline content [29].

Only a portion of the alkalis of the cement is readily soluble and strongly depends on the constituents of the clinker. Indeed, alkali sulfate are considered to be readily soluble. McCoy and Eshenour [117] showed that only 10 to 60 wt% of the total sodium and potassium content of the cement goes into solution after contact with water and even after one year, a considerable proportion of alkalis might not dissolve in the concrete pore solution. According to Taylor [164], the fractions of the rapidly water-soluble alkalis depend on the molar ratio of total SO_3 to total alkalis ($K_2O_{total} + Na_2O_{total}$). Moreover, the fraction of rapidly soluble K_2O is close to twice that of Na_2O, the latter tends more strongly to enter the silicate and aluminates phases. Alkalis less rapidly soluble could take place into silicate or aluminate phases as ionic substitution or in alkali sulfate inclusions in these phases [164]. The alkalis can also be bound during the cement hydration. In fact, a low Ca/Si ratio fosters alkali binding in calcium silicate hydrate (C-S-H) [62, 83].

Table 2.1 Typical alkali content of raw materials used in clinker manufacture (wt%) [88]

Alkali	Clay	Limestone	Shale	Chalk	Typical raw mix
K_2O (wt%)	2.61	0.26	4.56	0.04	0.52
Na_2O (wt%)	0.74	0.11	0.82	0.09	0.13

Table 2.2 Typical compositions of phases in Portland cement clinkers (wt%) (modified from Jawed and Skalny and Taylor [88, 164])

Clinker phase	Jawed and Skalny [88]		Taylor [164]	
	Na_2O (wt%)	K_2O (wt%)	Na_2O (wt%)	K_2O (wt%)
Alite (C_3S)	0.1–0.3	0.1–0.3	0.1	0.1
Belite (C_2S)	0.2–1.0	0.3–1.0	0.1	0.9
Aluminate (C_3A) (cubic)	0.3–1.7	0.4–1.1	1.0	0.7
Aluminate (C_3A) (orthorhombic)			0.6	4.0
Aluminate (C_3A) (low Fe)			0.4	0.5
Ferrite (C_4AF)	0.0–0.5	0.0–0.1	0.1	0.2
Ferrite (C_4AF) (low Al)			0.1	0.2

2.1.2 *Water*

Mixing water plays a major role in a number of concrete properties, either in its fresh state and, afterwards, in hardened concrete. Such influence comes from the chemical reaction between Portland cement (PC) and water (i.e., hydration), which brings non-negligible energy changes into the system. Therefore, potable water is considered as suitable to concrete production.

Depending on the source of the water, test procedures should be performed prior to its use in concrete such as pH along with sulfate, chloride, and alkali contents. ASTM C94/C94M [18] considers "distinct" types of potable water that may be used in concrete: mixing water (i.e., water for hydration); ice (water for controlling temperature rise); water for adjusting consistency (often added prior to pouring); free moisture within the aggregates and water within the chemical admixtures (when changes in the w/c ratio are greater than 0.01).

According to Neville and Brooks [165], water commonly used in concrete production presents an amount of dissolved solids that lies between 1000 and 2000 ppm, on average. The latter often displays inert characteristics and thus negligible influence on the properties of concrete.

As a rule, any water source with pH ranging from 6.0 and 8.0 which does not taste saline or brackish is suitable for concrete. Colour and odour do not necessarily mean that deleterious substances are present. Natural water sources that are slightly acidic are harmless, but water containing humic or other organic acids may adversely affect the mechanical properties of concrete [165].

Water usage must be in accordance with ASTM C1602/C1602M [16], which provides requirements, testing procedures and frequencies for qualifying individual or combined water sources. Table 2.3 compares the limits of impurities in mixing water, according to various standards. Methods of measuring solids in water and its quality are described in ASTM C114 [14] and ASTM C1603 [12].

Considering the above standards thresholds, to have an idea of the magnitude of alkalis provided by the water, one may consider the example of a conventional concrete with a water-to-cement ratio of 0.5 and a cement content of 420 kg/m^3. Thus, amounts of 0.126 kg/m^3, 0.210 kg/m^3 and 0.315 kg/m^3 would be provided respectively for ASTM C 1602, BS 3148 and BS EN 1008.

The evaluation of the importance of the alkalis contribution from the water should always be performed in combination with the other potential alkalis sources in the system. For example, BS EN 1008 [2] recommends that whether the water alkalis limit is exceeded (1500 mg/L), a given water source may be only used whether some primary prevention against ASR is adopted, such as the use of non-reactive aggregates, low alkali cement and/or supplementary cementitious materials (SCMs).

Nowadays, with the aim of reducing the carbon footprint in concrete construction, there is a greater demand for using by-products from different sources such as recycled materials and thus recycled water became a reality. In this case, different water sources other than potable water may be suitable for partial or complete usage in concrete. The latter includes, among others, groundwater, treated water from sewage, seawater, and reused water from ready mix concrete plants [102]. Special attention should be given to reused water from concrete ready-mix plants since high alkali concentrations and thus high pH may be found [109]. Hanna et al. [78] reported very high alkalinity (i.e. pH ranging from 12.5 to 13) with high concentrations of alkalis (i.e. Na_2O_{eq} 690 mg/L for concrete without chemical admixtures, 732 mg/L for water containing calcium lignosulfonate and 1005 mg/L for water containing naphthalene) in ready-mix reused water.

Shayan et al. [148] and Etxeberria et al. [65] studied the use of seawater in concrete. The authors reported important alkalis amounts in it, presenting 194 times greater amounts of Na ions when compared to potable water. Moreover, the high NaCl presence in the seawater was found to contribute to the early strength development in concrete. Yet, it may lead to durability related problems when added to concrete incorporating reactive aggregates and binders with high amount of aluminates [128].

Table 2.3 Optional chemical limits for combined mixing water

		BS 3148 [37]	BS EN 1008 [2]	ASTM C 1602/ C1602M [16]
Maximum concentration in combined mixing water, mg/L				
A. Chloride as Cl^-				
1	Prestressed concrete, bridge decks, or otherwise designated	–	500	500
2	Reinforced concrete in moist environments	500	1000	1000
3	Plain concrete	–	4500	–
B. Sulfate (SO_3)		1000	2000	3000
C. Alkalis as Na_2O_{eq} (Na_2O + 0.658 K_2O)		1000	1500	600

2.1.3 *Chemical Admixtures*

The use of chemical admixtures has significantly grown over the past decades to modify concrete properties in both fresh and hardened states. According to Mehta and Monteiro [119], approximately 90 wt% of all concrete produced in developed countries presents one or more types of chemical admixtures. However, engineers should be aware of the advantages and limitations of the use of chemical admixtures as well as the different chemical composition of each of them, which may vary from surfactants and soluble salts to polymers and insoluble minerals.

ASTM C125 [19] defines as a chemical admixture any material used as a concrete ingredient, added immediately before or during mixing and which is different from water, aggregates, hydraulic cement and fibres. The purpose of the use of chemical admixtures may vary from decrease costs of production to improve physical, chemical and/or mechanical properties of the material. ACI Committee 212 [9] lists a large number of different materials and their appropriate usage.

Some precaution should be taken in the use of chemical admixtures, particularly due to their chemical composition, which in some cases may induce undesirable effects especially regarding ASR, since they may bring important amounts of alkalis (Na^+ and K^+) to the system. Otherwise, according to Mehta [118] and Lindgård et al. [106] with the exception of accelerators, the latest generation of admixtures normally does not significantly contribute to the alkali loading of concrete. Although there is a number of chemical admixtures in the market with many distinct functions, compositions, and mechanisms, they may be divided in two large categories for further discussion: surface-active and set-controlling admixtures.

2.1.3.1 Surface-Active Admixtures

Surface-active admixtures, also known as surfactants, are generally used as air-entrainers or water reducers in concrete. According to Neville and Brooks [165], in what concerns air entraining action, surfactant ions are absorbed in the cement particles making them hydrophobic. Hence, the air bubbles generated during the concrete mixing are adhered to the cement paste. Some types of air-entrainers react with the calcium hydroxide to precipitate insoluble calcium salts. Air entrainers include vinsol resins, sodium adipates, sodium oleates and sodium lignosulphonates [40]. These admixtures normally contain in their chemical composition a total alkali content of approximately 10 wt%, which results in Na amounts of about 5.7 wt%.

Water reducers are chemical admixtures that act both chemically (i.e., electrostatic repulsion and steric hindrance) and physically (i.e. coating barrier) between the cement particles when added to the concrete mix. In general, low to mid- range water reducers are able to reduce the amount of water required down to 12 wt%, whereas for values above that, high range water reducers, the so-called superplasticizers are used [89, 134].

According to Ramachandran [136], water reducers of any kind may contain alkalis (i.e., Na and/or K) in their chemical composition. As a result, the alkalinity in the concrete pore solution may be raised and thus increase the likelihood of developing ASR in the presence of reactive aggregates. The amount of alkalis as Na_2O_{eq} of polycarboxylate-based superplasticizers usually ranges from 0.1 to 1.0 wt% whereas for naphthalene-based and melamine-lignin-based superplasticizers the values lie in between of 2.0–6.0 wt% [100, 172].

Leemann et al. [105] verified that the amount of sodium per cubic meter of concrete may increase on average 200 g with the use of superplasticizers. Hence, its influence on ASR development, even if it is small, cannot be neglected, especially because superplasticizers are conventionally used in high performance concrete where high strength and durability are initial assumptions of the material.

2.1.3.2 Set-Controlling Admixtures

According to Mehta and Monteiro [119] set-controlling admixtures can be divided into two groups: retarders and accelerators. Organic and inorganic compounds have been used as retarders. Organic compounds include unrefined Na, Ca, hydroxy-carboxylic acids (i.e., Na, Ca or triethanolamine salts of adipic, gluconic, tartaric, succinic, citric, and heptanoic acids), and carbohydrates. Inorganic compounds include oxides of Pb and Zn, phosphates, Mg salts, fluorates and borates [136].

Many inorganic compounds can also play the role of accelerators, including chlorides, fluorides, carbonates, silicates, aluminates, borates, nitrites, thiosulfates, etc., being calcium chloride one of the most efficient and economical accelerator used in conventional concrete [9, 119, 165]. Otherwise, accelerators used in shotcrete are normally different products, mainly based on aluminium silicates or salts [10]. Some commercial accelerators may contain high amounts of alkalis, which make their handling quite dangerous. Usually, their use results in a reduction of the long-term compressive strength and durability of the material.

Alkaline accelerators are composed mainly of sodium aluminate or sodium silicate. In the particular case of aluminate based accelerators, Na^+ and $(Al(OH)_4)^-)$ ions are incorporated in the cement paste to react with Ca^{2+} and SO_4^{2-} ions forming AFt (ettringite) and AFm (monosulfate) phases, which contribute to reduce the setting time of concrete [141, 184].

According to Myrdal [125], accelerators may contain up to 60 wt% of alkaline compounds in their chemical composition. Furthermore, NaSCN is probably the most recent non-chloride accelerator introduced in the market. Yet, its high amount of alkalis increases the potential to develop ASR in the presence of reactive aggregates and moisture. Another type of set-controlling admixtures is that controlling the freezing point of concrete that may contain sodium nitrite, sodium chloride and potassium carbonate, among others [140].

The demand for low alkali set-controllers is growing very rapidly in the construction market. They are normally water-based solutions of aluminium sulfate with a total alkali metal content (expressed as Na_2O_{eq}) lower than 1 wt%. However, considering the use of highly reactive aggregates, this amount of alkalis cannot be neglected. In order to develop new and high-performance set-controllers, several studies are currently in progress to explain their action mechanism and the factors affecting concrete setting [113]. Table 2.4 displays some of the most used concrete admixtures according to their classification, and alkali contents (i.e., Na_2O_{eq}).

2.1.4 Aggregates

Most of the types of rocks commonly used as aggregates contain minerals in which composition Na and/or K are present. Plagioclase, K-feldspar, K/Na-feldspar, feldspathoid, as well as some mica, clay minerals, zeolites and amphibole exist in a large number of aggregates. Also, dawsonite, volcanic glass and unwashed marine sands contain alkalis. Aggregates containing these minerals and materials have been investigated due to the suspicion that they might increase the alkalis content in concrete [33, 75–77, 95, 122, 123, 156, 157, 175, 177]. The release of the respective ions depends on the crystallographic structure of the minerals and on the bonding strength among the ions in the structure.

2.1.4.1 Solubility of Minerals

Some studies have shown that certain types of aggregates can release significant amounts of alkalis over time, when exposed to high pH environment. The rate and amount of alkalis released depends on generic exposure conditions including pH, Eh, temperature, local porosity and permeability of the possibly already cracked cement paste. In addition, alkali release is also affected by diverse properties of the actual aggregate, e.g., particle size, bulk lithological composition, mineral grain size, mineral modal content, total alkali content and bonding of mineral constituents related to the degree of alteration, and their spatial distribution in terms of microstructure, fabric or texture, as well as internal porosity and permeability of the aggregate particle [25, 34, 60, 110, 133].

Table 2.4 Concrete admixtures by classification and alkali content (i.e., Na_2O_{eq})

Type of admixture	Desired effect	Material	Na_2O_{eq} (wt%)
Accelerators	Accelerates setting and early-strength development	Calcium chloride (ASTM D98 and AASHTO M 144), triethanolamine, sodium thiocyanate, calcium formate, calcium nitrite, calcium nitrate	0–60
Air-entraining admixtures (ASTM C260/C260M and AASHTO M 154)	Improve durability against freeze–thaw, de-icer, sulfate, and alkali-reactive environments. Improve consistency of the mixture	Salts of wood resins (Vinsol resin), some synthetic detergents, salts of sulfonated lignin, salts of petroleum acids, salts of proteinaceous material, fatty and resinous acids and their salts, alkyl benzene sulfonates, salts of sulfonated hydrocarbons	7.0–10
Corrosion inhibitors	Reduce steel corrosion activity in a chloride-laden environment	Calcium nitrite, sodium nitrite, sodium benzoate, certain phosphates or fluosilicates, fluoaluminates, ester amines	0–5.0
Retarders (ASTM C494/C494M and AASHTO M 194M/ M 194, Type B)	Retard setting time	Lignin, Borax, sugars, tartaric acid and salts	0.1–1.0
Superplasticizers (ASTM C1017/ C1017M, Type 1)	Increase flowability of concrete for the same water-to-cement ratio	Sulfonated melamine formaldehyde condensates, sulfonated naphthalene formaldehyde condensates, lignosulfonates, polycarboxylates	2.0–6.0
Water reducers (ASTM C494/ C494M and AASHTO M 194M/ M 194, Type A)	Reduce water content down to 12 vol%	Lignosulfonates, hydroxylated carboxylic acids, carbohydrates	0.1–1.0

The groups of minerals that have been tested for alkali release by different authors are essentially the feldspars and the feldspathoids. The solubility of muscovite and biotite has also been analysed mainly with the aim to understand the behaviour of these minerals under variable temperature and pH conditions (e.g., [31, 47, 48, 112, 127, 131]).

Table 2.5 summarizes the most common minerals containing Na and/or K in rocks used as aggregates, their composition and crystallographic structure. Dawsonite, although rare, was included as it has been studied regarding alkalis release. Alkali ions (Na^+ and K^+) have weak bonds in minerals such as feldspars and mica mainly due to being monovalent when compared with bivalent Mg^{2+} and Ca^{2+} and therefore sodium is more easily exchangeable with potassium in the solution than the calcium, which has double bond in the mineral structure [107]. Another factor to consider is that the release of alkalis results also in the release of Si and Al to the solution, which are highly soluble at high pH.

Table 2.5 List of the most common minerals containing alkalis

Mineral family	Mineral name (variant)	Nominal composition	References
Framework silicates (tectosilicates)			
Alkali feldspar	Orthoclase (microcline, adularia)	$KAl[Si_3O_8]$	Ribbe [139]
	Anorthoclase	$(Na,K)Al[Si_3O_8]$	
Plagioclase	Albite (end-member)	$NaAl[Si_3O_8]$	
	Intermediate composition	$Na_{1-x}Ca_xAl[Si_{3-x}Al_xO_8]$	
	Anorthite (end-member)	$CaAl[Si_2AlO_8]$ (alkali-free)	
Zeolite	Analcime	$NaAlSi_2O_6{\cdot}H_2O$	Coombs et al. [46]
	Chabazite-Na	$(Na_3KCa_{0.5})[Al_{4.5}Si_{7.5}O_{24}]{\cdot}12H_2O$	
	Heulandite-Na	$(Na_4K_{0.5}Ca_2)[Al_8Si_{28}O_{72}]{\cdot}22H_2O$	
Feldspathoid	Nepheline	$(Na,K)AlSiO_4$	Deer et al. [53]
	Leucite	$KAlSi_2O_6$	
	Sodalite [group]	$Na_8Al_6Si_6O_{24}Cl_2$	
	Nosean	$Na_8Al_6Si_6O_{24}[SO_4]{\cdot}H_2O$	
	Hauÿne	$Na_6Ca_2Al_6Si_6O_{24}[SO_4]$	
	Lazurite	$Na_6Ca_2Al_6Si_6O_{24}S_2$	
	Cancrinite	$Na_6Ca_2Al_6Si_6O_{24}[CO_3]_2$	
Sheet silicates (phyllosilicates)			
Mica	Muscovite (phengite, sericite)	$KAl_2[Si_3AlO_{10}](OH)_2$	Rieder et al. [130]
	Illite [series name]	$K_{1-x}Al_2[Si_{3+x}Al_{1-x}O_{10}](OH)_2$	
	Paragonite	$NaAl_2[Si_3AlO_{10}](OH)_2$	
	Biotite [series name]	$K(Mg,Fe)_3[Si_3AlO_{10}](OH)_2$	
	Phlogopite	$KMg_3[Si_3AlO_{10}](OH)_2$	
	Glauconite [series name]	$(K,Na)_{<1}(Fe,Al,Mg)_2[Si,Al]_4O_{10}(OH)_2$	

(continued)

Table 2.5 (continued)

Mineral family	Mineral name (variant)	Nominal composition	References
Chain silicates (inosilicates)			
Amphibole (double-chain)	Hastingsite (colloquial: 'sodic hornblende')	$NaCa_2[(Fe,Mg)_4(Fe,Al)][Si_6Al_2O_{22}](OH)_2$	Hawthorne et al. [80]
	Glaucophane	$Na_2(Mg_3Al_2)[Si_8O_{22}](OH)_2$	
	Riebeckite	$Na_2(Fe_3Fe_2)[Si_8O_{22}](OH)_2$	
	Arfvedsonite	$NaNa_2(Fe_4Fe)[Si_8O_{22}](OH)_2$	
Pyroxene (single-chain)	Aegirine	$NaFe[Si_2O_6]$	Morimoto et al., [124]
	Jadeite	$NaAl[Si_2O_6]$	
	Aegirine-augite (colloquial: 'augite')	$(Ca,Na)(Mg,Fe)[(Si,Al)_2O_6]$	
	Omphacite	$(Ca,Na)(Mg,Al)[(Si,Al)_2O_6]$	
Non-silicates			
Sulfate	Alunite	$KAl_3[SO_4]_2(OH)_6$	Bayliss et al. [22]
Carbonate	Dawsonite	$NaAl[CO_3](OH)_2$	Stoica and Perez-Ramirez [161]
Fluoride	Halite	NaCl	
	Sylvite	KCl	

Notes
- Only selected alkali-containing minerals are listed here; many more exist
- Listed compositions are nominal: real-life compositions may deviate substantially
- Mineral names given are compliant with IMA-approved nomenclature per time of this writing (November 2022)
- Mineral names in tables and figures can be abbreviated cf. Whitney and Evans [180]

All feldspars consist of a three-dimensional network of (SiO_4) and (AlO_4) tetrahedra in which all tetrahedra share their oxygen atoms with the neighbours and Al, K, Na and Ca balance the charge between these tetrahedra [48]. Feldspars are grouped in two series, namely the alkali feldspars ($KAlSi_3O_8$–$NaAlSi_3O_8$) and the plagioclases ($NaAlSi_3O_8$–$CaAl_2Si_2O_8$), with the Si content decreasing from the Na- to the Ca- plagioclase. The crystal structures depend on the cations (Na, K and Ca) that are present, which have different sizes, with larger cations such as K^+ giving a monoclinic or nearly monoclinic symmetry, while the small cations, such as Na^+ and Ca^{2+}, cause a slight distortion of the structure and triclinic symmetry results. Even within the K-feldspar group there are some differences and, in the structure, and, according to Chao et al. [39], the difference between microcline and orthoclase is due to the atomic arrangement, assuming that in orthoclase the aluminium and silicon are randomly distributed, while in microcline these elements are considered to be in a particular set of lattice positions.

According to Vernon [176], in the alkali feldspar series, which forms a solid solution at high temperature, Na and K segregation takes place as temperature drops due to the size requirements of the crystallographic structure. The solid solutions between K- and Na-feldspars become metastable at temperature <600 °C, forming subparallel lamellae of alternately K- and Na-rich feldspar. These intergrowths are called perthites or antiperthites depending on the dominant type of feldspar.

The solubility of feldspars in water has been studied since 1930 [135]. The authors concluded that when finer ground fractions of feldspar are placed in water, decomposition takes place and alkalis are extracted from the feldspar. The alkalinity of the water is immediately increased and continues to increase with time although the rate of increase gradually becomes smaller with longer exposure.

K-feldspar has a slower absolute reaction rate than Na-plagioclase (albite) [183] and the dissolution rates of the feldspars in alkaline conditions depend on the concentration of the cations that can be involved in the cation-exchange and on the pH, as shown in the work of Blum [30]. However, albite seems to be more susceptible to be leached than K-feldspar in perthitic intergrowths when submitted to alkaline solutions as stated in Locati et al. [107] who concluded that the pre-existing microcracks and cleavages would influence the disaggregation of the minerals.

Phyllosilicate minerals have a structure in which the silicate tetrahedral form sheets [48]. Cations, which balance the charge, are positioned between the sheets. The study of the solubility of muscovite by Oelkers et al. [131] showed that alkali release at basic pH rates may be controlled by the breaking of octahedral Al–O bonds after adjoining tetrahedral Al and Si have been removed from the muscovite structure. The rates of solubility of muscovite, as well as in other aluminosilicates, decrease with increasing pH at acid conditions, minimize at near to neutral pH and increase with further pH increase at basic pH [131]. In addition, the experiments performed showed that at basic pH the dissolution rates of muscovite are dependent on aqueous Si and Al concentration. The effect of these concentrations on rates may be more significant that the effect of pH on these rates.

Minerals tested according to ASTM Chemical Method C289 (1N NaOH at 80 °C) have shown that biotite and quartz may release nearly identical concentrations of Si which is almost double of that of the feldspars [44]. For feldspars it has been proven that alkalis as well as Si and Al are released in significant quantities into alkaline solution. The work developed by Malmstrøm and Banwart [112] and Oelkers et al. [131] shows that dissolution depends on the ions already in solution as well as on temperature.

An important work developed by Wang et al. [178] on the study of the decomposition reactions of alkaline minerals in alkaline solution of pH > 13 indicate that the theoretical maximum alkali released occurs in the following order: dawsonite > leucite > nepheline > dehydration analcime > paragonite > muscovite > jadeite > Na-feldspar > K-feldspar. The authors also found that the alkalis release by minerals increase with increasing pH, except for nepheline, leucite and dawsonite.

The alkalis released from nepheline, K-feldspar and Na-feldspar depend on ion exchange between the surface layer of the minerals and Ca^{2+} in the solution [179]. The same authors found that nepheline is stable in solutions saturated in $Ca(OH)_2$ and decomposes when alkalis are added to the solution but K-feldspar and Na-feldspar are stable in alkaline solution, just showing some dissolution on the crystal interface. Locati et al. [107] studied the effect of alkaline solutions of 1N NaOH and KOH at 80 °C on K-feldspar and plagioclase. The polished thin sections of the minerals immersed in NaOH were observed by petrographic methods, namely optical microscope and scanning electron microscope and concluded that the attack was mainly evident along the weakness surfaces of cleavage planes, as also found by Ferraz et al. [66], twinning, and grain-grain contact as well as the finer phases in perthites. In consequence of the alkaline attack, Na_2O, K_2O, SiO_2 and Al_2O_3 were released by hydrolysis. The authors also found that crystal intergrowths both in K-feldspar and plagioclase are sensitive to alkali attack. Chou and Wollast [45] studied the weathering of albite at room temperature and concluded that the reaction involved three steps: instantaneous exchange of alkali ion by hydrogen ion; rapid build-up of a Na-depleted layer enriched in Si and/or Al; slow dissolution of the residual layer at the solid-solution interface accompanied by diffusion of ions from the fresh feldspar boundary. The first two steps occur mainly at dislocations or crystal defects. The long-term dissolution occurs in the steady-state third step.

According with Menéndez et al. [122] the extraction of alkalis from plagioclase and K-feldspars is observed in a concrete of a fifty years old dam with granodiorite aggregate. It is observed that there is a preferential release of minority alkali element in each feldspar, which is Na in K-feldspar and K in Na-feldspar. This is associated with the stability of the crystalline nets of the feldspar. Sodium migration occurs more easily for their ionic radius similar to the calcium.

In a recent work, Crundwell [47, 48] proposes a new theory of dissolution of silicates. In silicate minerals, Si forms a tetrahedron with four oxygen atoms that can be arranged in several different ways by sharing the oxygen atoms on the corners of the tetrahedral structure, resulting in the six classes of silicates: nesosilicates, sorosilicates, cyclosilicates, inosilicates, phyllosilicates, tectosilicates. Regarding the alkalis, the classes of interest are the tectosilicates, composed of a continuous framework of tetrahedral each sharing all four oxygen atoms (e.g., feldspars), the phyllosilicates, presenting continuous sheets of tetrahedral each sharing three oxygen atoms (e.g., mica) and inosilicates, in which the pyroxene group shows a continuous single chain of tetrahedral each sharing two oxygen atoms and the amphibole group contains a continuous double chain of tetrahedral alternately sharing two and three oxygen atoms. The order of reaction of the minerals with H^+ and with OH^- is presented in the same work, concluding that the clustering of the values obtained suggests that a general mechanism is responsible for the dissolution of all the silicate minerals [48]. The author also states that in alkaline solutions the dissolution of silicates follows two steps: (a) the metal atoms react with hydroxide ions form hydrolysed cations; and, (b) silicate groups react with water to form hydrated anions.

2.1.4.2 Influence of pH of Pore Solution

According with Yamada et al. [185] significant amounts of alkali can be supplied from clay minerals via an ion exchange with Ca from cement and from the dissolution of alkali rich minerals such as alkali feldspar or volcanic glass, etc. Alkali ions are released from aggregates, but it doesn't necessarily lead to an increase in pH, because alkalis are dissolved alkali silicate (or water glass). This reaction can reduce the available alkali and the water content, when these alkali silicate reacts with the $Ca(OH)_2$. Some of the alkali silicate or alkali silica gel is then extruded from the concrete, resulting in the reduction of the alkali and water.

2.1.5 Supplementary Cementitious Materials

Supplementary cementitious materials (SCMs) are industrial by-products that are used to substitute either Portland cement clinker in the production of composite cements or cement in the production of concrete. SCMs are materials that contribute to the properties of hardened concrete through hydraulic or pozzolanic activity. The most commonly used SCMs are fly ash, ground granulated blast-furnace slag and silica fume.

Fly ash is a fine particulate residue from the combustion of gases from pulverized coal in thermal power plant. The chemical composition of fly ash varies widely and depends on the types and relative amounts of non-combustible matter in the coal used [24]. For example, more than 100 different sources of fly ash produced in North America have been chemically analysed [149]; as a result, the alkali equivalent content (Na_2O_{eq}) of these materials ranges from less than 1.0 wt% to almost 10 wt%, for an average value of 2.44 wt%. Table 2.6 presents typical examples of chemical composition of fly ashes samples.

Ground granulated blast-furnace slag (GGBFS) is a glassy material formed when molten blast-furnace slag is rapidly chilled by immersing the molten slag in water, by the pelletizing process, or by other methods to ensure vitrification during the pig iron making process [101]. The chemical composition of a GGBFS varies depending on the composition of the raw materials, especially iron ore and coke, in the iron production process. However, since the raw materials are well controlled in the process, the composition of the GGBFS generally shows little variation within the same plant. Table 2.7 presents typical ranges of GGBFS composition.

Silica fume, also known as microsilica, is a finely divided residue resulting from the reduction of high-purity quartz with coal in an electric arc furnace in the manufacture of silicon or ferrosilicon alloy. Silica fume is characterized by its high amorphous silica content and its fine and spherically shaped particles less than 1 μm in diameter [101]. Silica fume is typically composed of amorphous silica in a range of 85 to 98 wt% SiO_2. Table 2.8 presents chemical composition of silica fume as reported by some authors.

Table 2.6 Typical chemical composition of fly ash samples (wt%) (Modified from Ramezanianpour [137])

Oxide (wt%)	1	2	3	4	5	6	7	8	9	10	11
SiO_2	47.1	44.1	35.5	38.3	45.1	48.0	55.7	55.6	55.6	46.3	44.5
Al_2O_3	23.0	21.4	12.5	12.8	22.2	21.5	20.4	23.1	21.4	22.1	21.1
$Fe_2O_{3\ total}$	20.4	26.8	44.7	39.7	15.7	10.6	4.61	3.48	2.99	3.10	3.38
CaO	1.21	1.95	1.89	4.49	3.77	6.72	10.7	12.3	11.0	13.3	12.9
MgO	1.17	0.99	0.63	0.43	0.91	0.96	1.53	1.21	1.76	3.11	3.10
K_2O	3.16	2.32	1.75	1.54	1.52	0.86	1.00	0.50	0.72	0.78	0.80
Na_2O	0.54	0.56	0.10	0.14	0.58	0.56	4.65	1.67	0.30	7.30	6.25
SO_3	0.67	0.96	0.75	1.34	1.40	0.52	0.38	0.30	0.16	0.80	7.81
SUM	97.25	99.08	97.82	98.74	91.18	89.72	98.97	98.16	93.93	96.79	99.84
Type of coal*	B	B	B	B	B	B	SB	SB	SB	L	L

*B = Bituminous coal; SB: Subbituminous; L: Lignite

Table 2.7 Typical ranges of GGBFS chemical composition (wt%) (Modified from Juenger et al. [91])

Oxide (wt%)	Min	Max	North America	Central and South America	Western Europe	Eastern Europe	India, Japan, Australia RSA
SiO_2	32	42	34.6–39.9	33.5–34.8	32.0–39.4	33.5–41.5	32.6–36.9
CaO	35	48	35.3–42.8	39.1–43.8	34.9–44.3	36.9–47.5	33.0–43.0
Al_2O_3	6	>19	6.6–11.5	10.0–13.0	9.5–12.5	5.5–12.4	10.2–19.3
MgO	3	14	7.0–13.1	5.9–9.9	5.0–13.4	2.5–11.2	4.9–13.8
TiO_2	0.2	>2	0.3–0.8	0.5–0.6	0.4–1.3	0.2–1.3	0.6–2.1
SO_3	1	4	2.0–3.0	1.1–3.7	2.0–4.5	1.6–3.8	1.7–4.0
Na_2O_{eq}	0.3	1.2	0.3–0.8	0.4–0.8	0.3–1.2	0.6–1.1	0.4–0.8

SCMs contain a significant amount of alkalis, particularly with respect to fly ash. In the case of certain SCMs, the alkali content is even significantly greater than that of Portland cement.

It is important to note that several SCMs, such as silica fume, blast furnace slag or fly ash, are used in concrete to reduce the effects of ASR. Indeed, it has been shown that these materials generally lead to a reduction in the concentration of alkali hydroxides in the concrete pore solution, which ultimately makes it possible to considerably reduce, to an acceptable level, the expansion experienced by the concrete when sufficient quantities of these materials are used [61]. This phenomenon can be explained by the chemical composition of these SCMs. Indeed, SCMs with the most effective behaviour contain a large amount of silica and low alkali content. SCMs rich in alkalis are generally less effective in controlling the alkali-silica reaction. An empirical relationship was proposed by Thomas [167] showing that the effectiveness of a fly ash to counteract the alkali-silica reaction is a function of the product of the

Table 2.8 Chemical composition of silica fume samples (wt%)

Oxide (wt%)	Yazici [186]	Maas et al. [111]	
SiO_2	92.26	94.28	94.69
Al_2O_3	0.89	0.04	0.21
$Fe_2O_{3\ total}$	1.97	0.06	0.12
CaO	0.49	0.51	0.71
MgO	0.96	0.57	0.29
K_2O	1.31	0.99	0.62
Na_2O	0.42	0.06	0.05
SO_3	0.33	0.16	0.10
LOI	n.a.	3.01	3.05
SUM	98.55	99.68	99.84

alkali-equivalent content and the calcium content divided by the square of the silica content $(Na_2O_{eq} \times CaO)/(SiO_2)^2$.

In Canada, CSA Standard A23.2-27A [155] dictates the requirements for SCMs and established “fly ash efficiency classes” for fly ash containing 0 to 3 wt% Na_2O_{eq}, 3 to 4.5 wt% Na_2O_{eq}, and >4.5 wt% Na_2O_{eq}. The same document also states that the silica fume content needs to counteract the alkali-silica reactivity and will need to be modulated based on the alkali content of the concrete (calculated in kg/m^3). In addition, the silica fume and ground granulated blast furnace slag must contain less than 1 wt% alkali (Na_2O_{eq}) to be used in the presence of alkali-silica reactive aggregates. If their alkali content is greater than 1 wt% Na_2O_{eq}, then their effectiveness must be demonstrated by laboratory expansion tests. Thus, although SCMs contain alkalis, some of them still make it possible to reduce the alkalinity of the pore solution of the concrete by their great capacity to incorporate alkalis into the structure of C-S-H [61, 167].

After the hydration of the cement with supplementary cementitious materials part of the alkalis remain in these SCMs. A part of these alkalis can release over a long time period and contribute to increase the alkali content in the pore solution of the concrete. But, in general, a moderate potential of extraction of alkalis from SCMs with the time is observed. This extraction varies with respect to the type of SCMs, the percentage of extraction with the time is between 0.02 and 0.04 wt% in most of the fly ash, but the slags show extraction of alkalis of up to 0.20 to 0.25 wt% or even higher percentages. In natural pozzolans 0.30 to 0.40 wt% are extracted [120].

According to Yamada et al. [185], SCMs are included in some cements, and contain various amounts of alkalis as water-soluble minerals and dissolved in glass phases. With hydration or pozzolanic reactions, the hydration products have lower C/(S + A) than Portland cement. This reaction consume water, but the hydration products with the lower C/(S + A) takes up alkalis depending on the level of C/(S + A). The C/(S + A) is determined by both the Ca content and the degree of hydration of SCMs.

2.1.6 Recycling Alkalis by Alkali-Silica Reaction Products

Bérubé et al. [28] suggested that monitoring the depletion of calcium hydroxide in the paste could be a useful complement to measuring alkalis. If alkali recycling is occurring, calcium hydroxide acts as a source of calcium that replaces sodium and/or potassium in the ASR gel. Even though the sodium and potassium content of the pore solution may remain stable, the amount of solid calcium hydroxide will decrease. Because solid calcium hydroxide is crystalline, it can be measured using X-ray diffraction (XRD). The downside to this technique is that sample preparation for XRD is essentially a destructive process, preventing repeated measurement over time of the calcium hydroxide content of a specimen.

When the alkali-silica products of reaction are observed by electron microscopy combined with X-ray microanalysis, it is observed that the first products formed are gels. In addition, microcrystalline products in the first steps have a high concentration of sodium (needle like crystals) and these are evolving to other microstructures with less quantity of sodium and high proportions of calcium (rosette like crystals). This is associated with the modification of composition of the ASR products with release of sodium and incorporation of low quantities of potassium and calcium [49, 121].

According to Katayama et al. [92, 93], the compositions of ASR gels (crack-filling within aggregate, crack-filling within cement paste, air voids) and their crystalline products (rosette-like crystals) can be determined by energy dispersive X-ray spectroscopy (EDS). During the migration of ASR gel along cracks from the reacted aggregate into cement paste, this gel absorbs Ca ions from cement paste and loses alkalis, approaching "convergent points" at which apparent chemical equilibrium is reached with C-S-H gel in the cement paste, whereby losing expansion potential.

2.1.7 Other Components (Fibres)

To improve the properties of the concrete, different types of fibres are used. Some of the properties related with the fibres are: mechanical strength, tensile strength, fire resistance, avoidance of crack formation and creep. Some of these fibres have alkalis in their composition and theoretically can increase the content of alkalis in the concrete. However, increase of alkali-silica reaction due to presence of fibres hasn´t been reported.

- *Fibres with alkalis in their composition*

The fibres that have alkalis in their composition are mainly glass fibres. There are different types of glass fibres with different properties and chemical compositions.

According [79] there exist high strength glass fibres that are described in Table 2.9 [20]. These types of glass were developed to provide combinations of fibre properties directed at specific end use applications. Their chemical compositions are described in Table 2.10.

The content of alkalis varies for different types of fibres from very low percentages to values higher than 10 wt%.

Table 2.9 Types of glass fibres [20]

Type	Description
A Glass	Soda lime silicate glasses used where the strength, durability, and good electrical resistivity of E Glass are not required
C Glass	Calcium borosilicate glasses used for their chemical stability in corrosive acid environments
D Glass	Borosilicate glasses with a low dielectric constant for electrical applications
E Glass	Alumina-calcium-borosilicate glasses with a maximum alkali content of 2 wt% used as general purpose fibres where strength and high electrical resistivity are required
ECR® Glass	Calcium aluminosilicate glasses with a maximum alkali content of 2 wt% used where strength, electrical resistivity, and acid corrosion resistance are desired
AR Glass	Alkali resistant glasses composed of alkali zirconium silicates used in cement substrates and concrete
R Glass	Calcium aluminosilicate glasses used for reinforcement where added strength and acid corrosion resistance are required
S-2® Glass	Magnesium aluminosilicate glasses used for textile substrates or reinforcement in composite structural applications which require high strength, modulus, and stability under extreme temperature and corrosive environments

Table 2.10 Chemical composition (wt%) of glass fibres [79]

Oxide	Glass type							
	A	C	D	E	ECR®	AR	R	S-2®
$Na_2O + K_2O$	14–16	7–10	0–4	0–2	0–2	11–21	0–1	0–0.2
MgO	0–4	2–4	–	0–5	0–4	–	4–7	9.5–10
CaO	6–10	11–15	0–1	16–25	17–25	1–10	8–15	0–0.2
Fe_2O_3 total	0–0.5	0–0.8	0–0.3	0–0.8	0–0.8	0–5	0–0.5	0–0.1
B_2O_3	0–6	4–6	21–24	5–10	–	0–8	0–0.35	–
Al_2O_3	0–6	3–5	0–1	12–16	9–15	0–5	23–28	24–25
SiO_2	63–72	64–68	72–75	52–56	54–62	55–75	55–60	64–66
TiO_2	0–0.6	–	–	0–1.5	0–4	0–12	–	–
Other	F 0–0.4	BaO 0–1		F 0–1	ZnO 2–5	Li_2O 0–1.5 ZrO_2 1–18 F 0–5	F 0–0.3	

- ***Fibres without alkalis in their composition***

There also exist fibres without alkali content in its chemical composition such as synthetic and metallic fibres that are used to improve the properties of the concrete. The types and alkali content are shown in Table 2.11 and Table 2.12, respectively [68, 170, 171].

Table 2.11 Type of synthetics fibres [171]

Fibre types	Acrylic	Aramid	Carbon	Polypropylene	Polystyrene	Nylon	Polyester
Alkali content	Alkali-free	Alkali-free	Alkali-free	Alkali-free	Alkali-free	Alkali-free	Alkali-free

Table 2.12 Type of metallic fibres [68, 170]

Fibre types	Description	Alkali content
Carbon steel	Iron, carbon: 1 wt% Silicon, manganese, sulfur, phosphorous, oxygen, hydrogen	Alkali-free
Stainless steel	Chromium: 10–12 wt% Molybdenum, nickel	Alkali-free

2.2 External Sources

The importance of considering potential external sources of alkali metal ions when predicting the residual reactivity and thus the future performance of ASR affected structures was, among others, addressed in Thaulow and Geiker [166].

2.2.1 Alkaline Water

Water can be naturally alkaline when passing through rocks or minerals that could transfer some basic elements. For instance, the natural water of Monchique, which is located in the Portuguese Algarve, is considered the most alkaline drinking water in the world with a pH of 9.5 [86].

Also, there are medicinal waters called "sodium-rich waters" that are used in spas because they have numerous health benefits, according to [23]. There are many locations throughout the world with sodium-rich waters. For example, in several areas of Spain, on the northern coast of Bulgaria by the Black Sea, on the northern coast of Poland next to the Baltic Sea and in the vicinity of the Sudetenland, in western Germany in the Nahe River valley, in the south and centre of France, as well as in the northeast of Slovenia in the border with Austria.

There is also sodium-rich water in some points of the northern coast of Morocco, south of Qatar on the border with Saudi Arabia, or for instance on the west coast of the United States, specifically in San Luis Obispo (California) and south of Canada in the Saskatchewan region.

2.2.2 Sea Water

Seawater, or salt water, is water from a sea or ocean. On average, seawater in the world's oceans has a salinity of about 3.5 wt% (35 g/L). This means that every kilogram (roughly one litre by volume) of seawater has approximately 35 g of dissolved salts (predominantly sodium (Na^+) and chloride (Cl^-) ions). The average density at the surface is 1.025 kg/L. Seawater is denser than both fresh water and pure water (density 1.0 kg/L at 4 °C) because the dissolved salts increase the mass by a larger proportion than the volume. The freezing point of seawater decreases as salt concentration increases. At typical salinity, it freezes at about −2 °C [132]. The coldest seawater ever recorded (in a liquid state) was in 2010, in a stream under an Antarctic glacier, and measured −2.6 °C [174]. Seawater pH is typically limited to a range between 7.5 and 8.4 [43].

Figure 2.1 shows schematically the concentrations of various salt ions in seawater.

Dissolved salts in seawater are electrolytes in ionized aqueous solution, which gives seawater complex physical–chemical properties. Table 2.13 shows the most important salts found in seawater obtained from the course "geography of the sea" [38].

In the Fig. 2.2 the proportion of salts in seawater is shown.

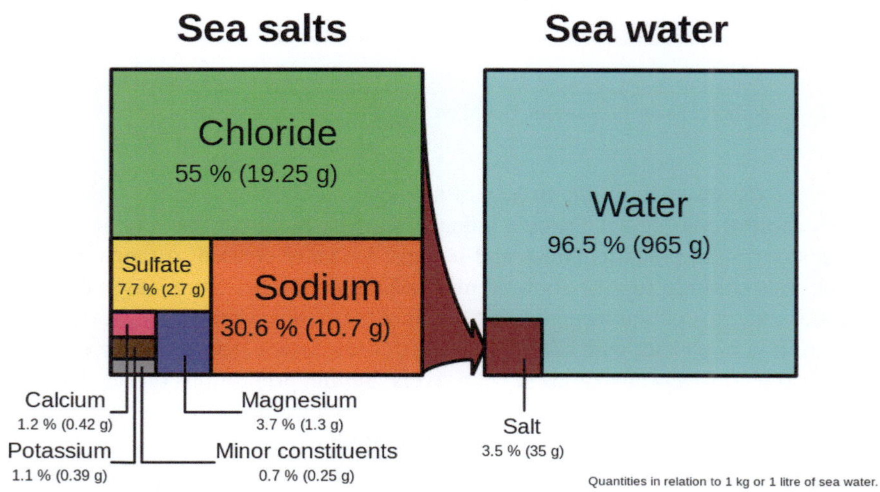

Fig. 2.1 The relative proportions of ions in seawater. (By derivative work: Tcncv (talk) Sea_salt-e_hg.svg: Hannes Grobe, Alfred Wegener Institute for Polar and Marine Research, Bremerhaven, Germany; SVG version by Stefan Majewsky (Sea_salt-e_hg.svg) [CC BY-SA 2.5], via Wikimedia Commons)

Table 2.13 Amounts and proportions of dissolved salts in average sea water of salinity S = 35, in grams per litre (g/L), from UC [38]

Salt	Composition	Content (g/L)	Total (wt%)
Sodium chloride	$NaCl$	27.213	77.7
Magnesium chloride	$MgCl_2$	3.807	10.9
Magnesium sulfate	$MgSO_4$	1.658	4.74
Calcium sulfate	$CaSO_4$	1.260	3.60
Potassium sulfate	K_2SO_4	0.863	2.47
Calcium carbonate	$CaCO_3$	0.123	0.35
Magnesium bromide	$MgBr_2$	0.076	0.22
SUM	–	35.000	99.98

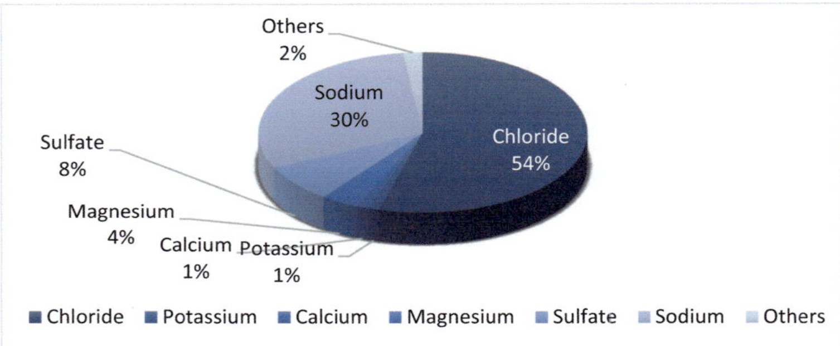

Fig. 2.2 Proportion of salts in seawater. Based on UC [38]

Most of the research on the impact of seawater on concrete focuses on chloride ingress and the consequent chloride induced reinforcement corrosion. However, as can be seen in the previous tables and images, seawater contains a range of other ions besides chloride such as sodium, magnesium, sulfate, calcium and potassium. Seawater generally acts as an endless reservoir of these ions compared to the concrete structure it is in contact with. The total salt concentration in seawater can vary, for example due to freshwater dilution from rivers, but the proportions of the elements are generally similar. In the following Table 2.14 is shown the seawater composition from the Atlantic, the Trondheim Fjord and the Baltic Sea.

In a study carried out by Menéndez [120], water from the Atlantic Ocean has been analysed, specifically samples taken from Fuerteventura, using the XRF and ICP techniques, obtaining the following composition shown in Table 2.15 and Table 2.16, respectively.

The dry residue of the seawater sample was determined following the method described in Standard UNE 83,957:2008. The result obtained is: 50.26 mg/kg.

Based on a data set comprising 21 different concretes exposed (mainly submerged) to sea water in 9 different locations with exposure times ranging from 2 to 34 years, a

Table 2.14 Concentrations of the main elements in sea water from the Atlantic Ocean [54], Trondheim Fjord and the Baltic Sea [81]. The results are expressed both in mg/kg and in percentages

	Cl^-	Na^+	Mg^+	SO_4^{2-}	Ca^{2+}	K^+
Concentration (mg/kg)	3960–21000	2190–11100	260–1410	580–2970	50–480	70–400
Concentration (wt%)	55.25–55.70	29.62–30.80	3.58–3.90	8.00–8.15	0.7–1.33	0.98–1.11

Table 2.15 Concentrations of the main elements in sea water from the Atlantic Ocean obtained using XRF [120]

Conc. (g/L)	Al	Ca	Fe	K	Mg	Na	S	Si	P	Cl	Br	Mo	Cs
Atlantic seawater	0.00018	1.69	0.03	1.19	4.07	32.85	2.72	0.29	1.31	55.25	0.23	0.04	0.31

Table 2.16 An example of the total alkali content in an Atlantic water sample analysed by ICP [120]

Concentration (g/L)	Na_2O	K_2O	Na_2O_{eq}
Atlantic seawater	44.28	1.44	45.22

typical elemental zonation near the exposed concrete surface was observed irrespective of the age, location and binder composition [87]. Figure 2.3 gives a schematic overview of the different zones with the corresponding phase changes and elemental profiles near the seawater-exposed concrete surface. Besides a calcite and brucite containing crust, three distinct zones were observed. The outermost zone of the concrete is heavily leached in amongst others calcium, and contains instead magnesium rich phases as brucite and magnesium silicate hydrate (M-S-H) [50]. The magnesium enriched zone is followed by a sulfate enriched zone where gypsum, ettringite and for low temperatures and in the presence of carbonates even thaumasite can be observed [90, 152]. The sulfate enriched zone slightly overlaps with the start of the chloride enriched zone where chlorides can be bound in for example calcium chloroaluminate hydrates, such as Friedel's salt or Kuzel's salt, or be adsorbed onto calcium silicate hydrate (C-S-H). The zonation is associated with the differences in the mobility of the ions and the interactions between the ions and the solids. The magnesium and sulfate enriched zone for the marine concretes investigated in [87] did not reach further in than 10 mm from the exposed surface whereas chlorides show clear ingress profiles up to 100 mm into the concrete.

Even though the sodium concentration in for example the Atlantic Ocean is about 11495 mg/kg whereas the pore solution merely contains about 3448 mmol/L [52], only a limited amount of sodium is observed to move into the concrete by diffusion, as illustrated in Fig. 2.4.

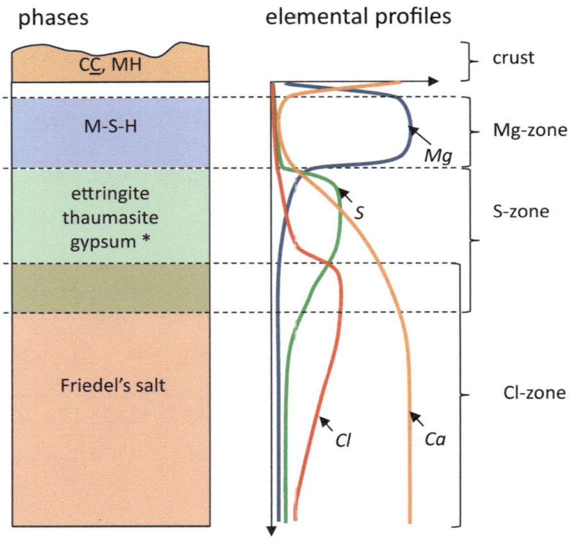

Fig. 2.3 Schematic overview of the phases and elemental profiles observed in seawater exposed concrete (extracted from Jakobsen et al. [87])

Sodium does not follow the chloride into the concrete. The reason why we do not obtain sodium profiles in seawater exposed concrete, is the limited binding capacity of the cement hydrates for sodium compared with the, for example, large affinity of the hydrates to chloride. One could actually say that cement hydrates have a built-in indicator for the ingress of chlorides as the interaction between the chloride and the solids allows us to detect the chlorides and track their ingress into the concrete. This is not the case for sodium. Only the outermost heavily leached zone of the concrete shows an enrichment in sodium, shown in Fig. 2.5, indicating that the heavily leached solids can sorb sodium.

Potassium levels in the seawater are low compared with the concentration in the pore solution, for example, about 391 mg/kg in seawater versus 7820 mg/kg in pore solution [52]. Potassium therefore leaches out of the concrete upon submerged exposure to seawater, as demonstrated in the elemental profiles shown in Fig. 2.5 from De Weerdt et al. [51, 52].

In conclusion, submerged exposure to seawater leads to a clear enrichment in carbonates, magnesium, and sulfates in the outer 10 mm and enrichment in chlorides deeper in the concrete. With regard to alkali metals, submerged exposure to seawater does not seem to lead to a considerable enrichment of sodium in the concrete, but leads to a lowering of the potassium content due to leaching. Submerged exposure to seawater has thus a lowering effect on the alkali metal content in concrete and might thereby reduce the risk for ASR in the leached zone.

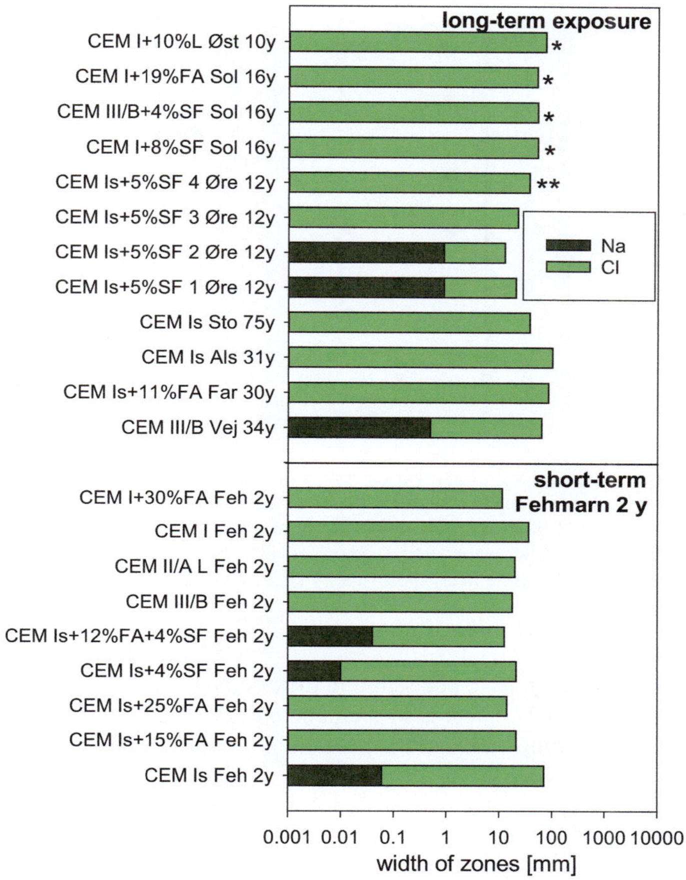

Fig. 2.4 Bar diagrams comparing the ingress depth of Na with that of Cl in the concretes tested in Jakobsen et al. [87]. The notation refers to the composition of the binder, the exposure site and the exposure time; for more information see Jakobsen et al. [87]. (*) indicates that the end of the chloride-rich zone was determined by titration; (**) indicates high chloride at the end of the section (based on Jakobsen et al. [87])

2.2.3 *Fertilizers*

These are formed by two or more main nutrients (nitrogen, phosphorus and potassium) and may contain some of the three secondary nutrients (calcium, magnesium, and sulfur) or micronutrients (boron, copper, iron, manganese, molybdenum and zinc) essential for the growth of plants, although in small quantities compared with the main and secondary nutrients [4]. The main types of fertilizers are shown in the Table 2.17.

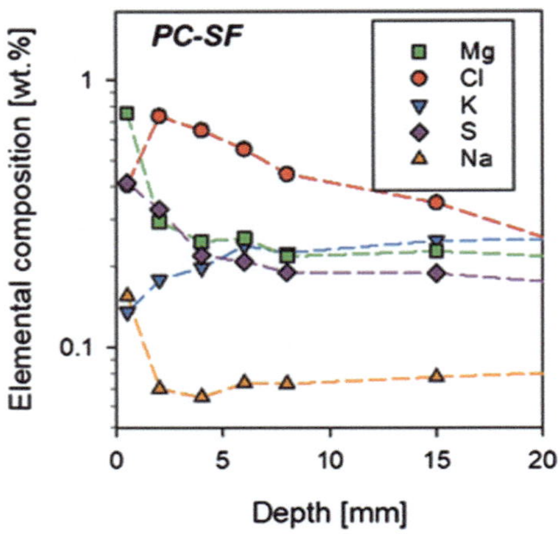

Fig. 2.5 Changes in Cl, K, Mg, S and Na content as a function of the depth from the exposed surface for concrete beams prepared with Portland cement and 8 wt% silica fume (PC-SF) and a water to binder ratio of 0.44 exposed submerged in seawater at the west coast of Norway for 16 years. The elemental profiles were determined on acid dissolved profile ground concrete powder using inductively coupled plasma mass spectrometry (ICP-MS). (Modified from De Weerdt et al. [51])

Some of the fertilizers have potassium in their composition. These alkalis can be leached with the water and access to the concrete, increasing or the risk of ASR. Below are the main components of the potassium fertilizers collected in Table 2.18 [114].

2.2.4 *Industrial Wastes*

The wastewaters from industrial processes have very different compositions, depending on the industrial process in general and on the specific process in different factories. In relation with the alkaline characteristics of these waters, Bond and Straub [32], give values of pH from different industrial processes.

In Table 2.19 the characteristics of these waters are shown. It is observed that industries like cotton, wool, tannery and potato processing or coke oven can produce highly alkaline waters.

Compliance with current environmental requirements makes it necessary for alkaline industrial wastewater to be treated to reduce its pH to a sustainable level in the different points of final disposal.

Carbon dioxide (CO_2) is an inert, colourless and non-toxic gas, which in presence of water forms carbonic acid allowing the neutralization of alkaline liquid streams. This constitutes an alternative method and a potential substitute of inorganic acids (sulfuric and hydrochloric acid) which are traditionally used in the industry, obtaining numerous benefits with its application [84].

Table 2.17 Composite fertilisers [4, 85]

Fertilisers	Subtype	Compounds	Additional description
Binary or double fertilisers	NP	$(NH_4)H_2PO_4$ $(NH_4)_2HPO_4$	
	NK	K_3PO_4 K_2HPO_4	
Double deficiency correctors	–	CaB CaMg	Fertilisers to correct the nutritional deficiency of 2 specific nutrients that are usually related.
Ternary or triple fertilizers	NPK (main nutrients of plants)	N P_2O_5 K_2O	Industrially produced chemical fertilizers contain a guaranteed minimum quantity of nutrients, and are indicated on the bag.
Triple deficiency correctors	–	Fe Mn Zn	Fertilizers to correct the nutritional deficiency of 3 certain nutrients that are usually related or whose deficiencies are difficult to discern between them.
Multicarencial correctors	–		Fertilisers to correct more than 3 nutritional deficiencies.
Potassium nitrate	–	KNO_3 13 wt% N (58 wt% NO_3) 38 wt% K (46 wt% K_2O)	
Potassium sulfate	–	K_2SO_4 43% K (52 wt% K_2O) 18% S (54 wt% SO_4)	

2.2.5 *De-Icing/Anti-Icing Salts*

For a long time, it has been assumed that external alkalis from de-icing/anti-icing salts can trigger and/or accelerate ASR in concrete. The salts are either applied to melt accumulated layers of snow and ice on the pavement surface (de-icing) or they are applied prior to an expected precipitation event on the bare pavement surface (anti-icing) to prevent the accumulation of snow and ice. Below, only the more common term “de-icing” will be used, but always both procedures (de-icing/anti-icing) are addressed.

The research in the last decades mainly focused on sodium chloride (NaCl), the most frequently used de-icing salt in the world for pavements, first used in Paris in 1880 [63]. The applied quantity of NaCl on concrete pavements highly depends on the weather situation and can vary between 5 and more than 60 g/m^2 with every application. In Germany, for example, between 10 and 40 t of NaCl are applied per km and winter on highway pavements [69]. For the application, NaCl (70 wt%)

Table 2.18 Chemical composition of potassium fertilisers [114]

Product	Nitrogen (wt% N)	Phosphorus (wt% P_2O_5)	Potassium (wt% K_2O)	Sulfur (wt% S)	Magnesium (wt% Mg)
Potassium chloride			60–62		
Potassium nitrate	13–14		44–46		
Potassium sulfate			50–53	18	
Potassium and magnesium sulfate			22	22	11
Potassium carbonate			27–52		
Potassium pentaborate			13.3		
Potassium monophosphate		51	35		
Potassium diphosphate		41	54		
Potassium polyphosphate		51	40		
Potassium thiosulfate			25	17	

is usually pre-wetted with a 20 vol% solution of NaCl (30 wt%), sometimes also of $CaCl_2$ or $MgCl_2$, to achieve a better adhesion to the pavement surface and to reduce a wind-blown dispersion [63, 99]. Many applications are done shortly before freezing is expected, i.e., NaCl is applied directly onto the concrete surface. For safety reasons, chloride-based de-icers are not used for airfields in order to avoid corrosion damage to aircraft and airport equipment. Therefore, synthetic urea (CH_4N_2O) was used as airfield de-icer in the past. But due to environmental concerns because of the N-input in the surrounding soil and only short action times, new de-icers based on sodium and potassium acetate (CH_3COOK/Na) as well as sodium and potassium formate (HCOOK/Na) were introduced in the early 1990s. These de-icers exhibit a large number of advantages over the synthetic urea. They are easily bio-degradable, leaving behind only harmless, non-toxic residues and contain no nitrogen. They are also more effective at lower temperatures (up to −60 °C), work faster and longer and require fewer applications. The K-based de-icers are 50 vol% solutions of CH_3COOK or HCOOK and are either applied alone or in case of an ice or packed snow layer together with the Na-based de-icers (granules) for a higher efficiency. The application rates usually range between 10 and 60 g/m^2. In Germany, for example, there are approximately 40–50 days of de-icing within every winter and, if necessary, up to eight applications are done per day. In total, between 400 and 2000 t or even more of these de-icers are applied per airport and winter [181].

Table 2.19 Comparative strengths of wastewater from industry [32]

Type of waste	BOD (mg/L)	COD (mg/L)	SS (mg/L)	pH
Apparel				
Cotton	200–1000	400–1800	200	8–12
Wool scouring	2000–5000	2000–5000[a]	3000–30000	9–11
Wool composite	1	–	100	9–10
Tannery	1000–2000	2000–4000	2000–3000	11–12
Laundry	1600	2700	250–500	8–9
Food				
Brewery	850	1700	90	4–8
Distillery	7	10	Low	–
Dairy	600–1000	150-250[a]	200–400	Acid
Cannery				
Citrus	2000	–	7000	Acid
Pea	570	–	130	Acid
Slaughterhouse	1500–2500	200-400[a]	800	7
Potato processing	2000	3500	2500	11–13
Sugar beet	450–2000	600–3000	800–1500	7–8
Farm	1000–2000	500–1000[a]	1500–3000	7.5–8.5
Poultry	500–800	600–1050	450–800	6.5–9
Materials				
Pulp; sulfite	1400–1700	84–10000	Variable	
Pulp; kraft	100–350	170–600	75–300	7–9.5
Paperboard	100–450	300–1400	40–100	
Strawboard	950	850[a]	1350	
Coke oven	780	1650[a]	70	7–11
Oil refinery	100–500	150–800	130–600	2–6

[a] = CODMn, mg O/L

2.2.5.1 The Influence of Sodium Chloride on ASR

In the late 1990s an increasing number of ASR-damaged concrete highway pavements occurred in Germany although the existing guideline for preventing ASR in concrete, first released in 1974, was strictly followed [67]. Affected highway sections showed first signs of ASR after about 8–15 years in service, i.e., after half the intended service life of 30 years at the latest. The high amount of highway pavement concrete (nearly 50 wt%) on the total number of ASR-damaged structures led to the suspicion that special mechanisms promote the ASR in pavements. First studies (e.g., [144, 145, 158, 158] revealed that an application of NaCl de-icer solution on concrete prisms with alkali-reactive aggregates significantly accelerates the ASR (Fig. 2.6).

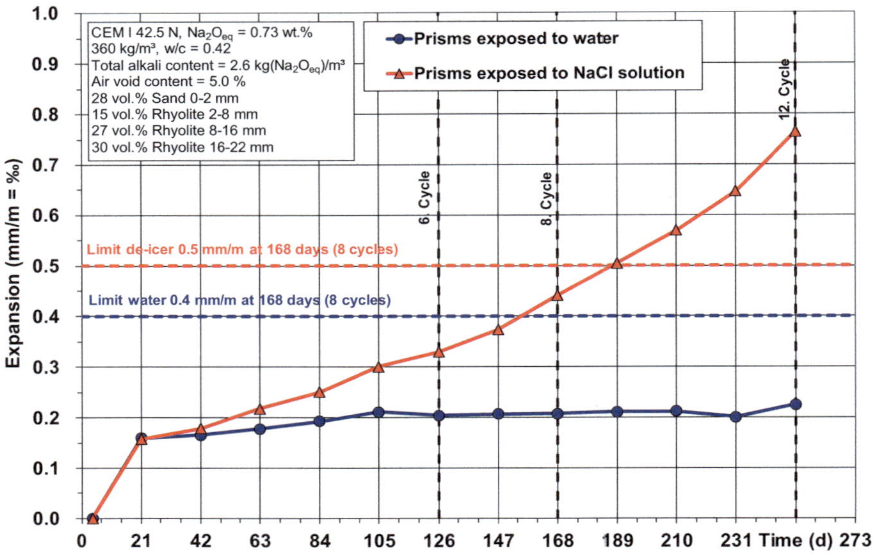

Fig. 2.6 Influence of NaCl solution on expansion during the ASR performance test (CS-CPT)

The influence of NaCl on ASR has been studied frequently since the 1970s. Locher and Sprung suggested that NaCl can only be active for an ASR after a transformation into NaOH, for what the Cl^- ions need to be fixed in cement hydration products [108]. Chatterji et al. [42] showed that mortar bars gave much higher expansions when stored in solutions of alkali salts than in NaOH solution, suggesting a direct influence of the alkali salts not related to a transformation into NaOH [42]. It was concluded that in contrast to NaOH, NaCl does not lower the Ca^{2+} concentration in the pore solution and the higher the Ca^{2+} concentration the lower the diffusion of dissolved silica out of the aggregate grains what consequently results in higher expansion pressures [41]. A series of studies from Kawamura et al. clearly showed the accelerating effect of NaCl on ASR and linked the mechanism to a direct influence of the Cl^- ions [94, 96–98]. Bérubé and Frenette [26] found that NaCl will only exacerbate ASR in concrete with a high alkali content, while for low-alkali cements NaOH is more severe. It was concluded later that low-alkali cements are sufficient to avoid ASR in concrete exposed to NaCl, but the role of expansive chloroaluminates should be studied further [27]. Sibbick and Page concluded that a supply of NaCl to concrete results in a formation of Friedel's salt ($C_3A.CaCl_2.10H_2O$) which increases the NaOH concentration in the pore solution and finally accelerates the ASR, also in concrete with a low alkali content [150, 151]. Shayan [146] found that a higher expansion for concrete exposed to NaCl solution is related to expansive chloroaluminates and ettringite rather than to ASR. Later, Brown and Bothe [35] suggested that the formation of calcium oxychlorides results in an increase of the pH accelerating the ASR what could not be confirmed, however, in the study from Hooton and

Julio-Betancourt [82]. Overall, most of the studies showed that an external supply of NaCl can trigger and accelerate ASR in concrete but there has been no consensus on the mechanism.

Recent studies [70–72] showed that an addition of NaCl to concrete pore solutions does not result in an increase of the pH. The findings show that in the presence of NaCl, the dissolution of silica is promoted directly, beyond the influence of the pH (Fig. 2.7). This is in agreement with the studies from Chatterji et al. [42] and especially from Kawamura et al. [94, 96] who had already suggested a more "direct" influence of alkali salts on ASR. Further support can be found in studies from Dove and Elston [56] showing a significant increase of the silica dissolution rate in NaCl solutions, even for a pH below 12. The mechanism assumed is a formation of $\equiv SiO^{-}Na^{+}$ complexes at the silica surface resulting in higher silica dissolution rates [56]. Similarly, Zhang and Liu concluded recently that the adsorption of alkali metal ions as Na^{+} on the silica surface weakens the Si–O bond and accelerates the hydrolysis reaction [187]. In the studies by Dressler et al., the formation of an aqueous $NaHSiO_3$ complex was proposed that again would result in an increase of the silica solubility [58, 59]. However, the more severe impact of NaCl compared to NaOH as often reported (e.g., [42, 96] cannot be explained by a solely Na-driven mechanism. The anion (Cl^{-}), as already assumed by Kawamura et al. [96] seems to play a role too. A mechanism as the salting-in/out effect [115, 162, 163] appears to be more likely. Further research is in progress to study this mechanism in detail. The formation of Friedel's salt could be confirmed as reported earlier [27, 146, 150, 151] but is related to the transformation of ettringite. No additional OH^{-} ions will be released:

$$C_3A \cdot 3CaSO_4 \cdot 32H_2O + 2Na^{+} + 2Cl^{-} \leftrightarrow C_3A \cdot CaCl_2 \cdot 10H_2O + 2Ca^{2+} + 2Na^{+} + 3SO_4^{2-} + 22H_2O$$

Only a formation of Friedel's salt from C_3A, C_4AF and alite would increase the pH because of the accompanying dissolution of portlandite:

$$C_3A + 2NaCl + Ca(OH)_2 + 10H_2O \leftrightarrow C_3A \cdot CaCl_2 \cdot 10H_2O + 2NaOH$$

This happens if NaCl is added to the mixing water, as shown in earlier studies [129]. Hence, different mechanisms run when NaCl is added to the mixing water or when it is applied externally to hardened concrete. The vital role of $Ca(OH)_2$ is to resupply OH^{-} ions that are consumed by the dissolution of silica and to provide Ca^{2+} ions to form ASR products. In concrete with low-alkali Portland cement (CEM I) the lower pH may slow down the dissolution of alkali-reactive silica, but the application of NaCl will accelerate its dissolution beyond the effect of the pH and may cause deleterious ASR in the long term. Thus, ASR in concrete exposed to NaCl may be delayed but not prevented permanently when low-alkali cement (CEM I) is used. The duration of the delay depends on the alkali content of the cement too, but primarily on the alkali-reactivity of the aggregates and the intensity, i.e., amount and concentration of the externally applied NaCl.

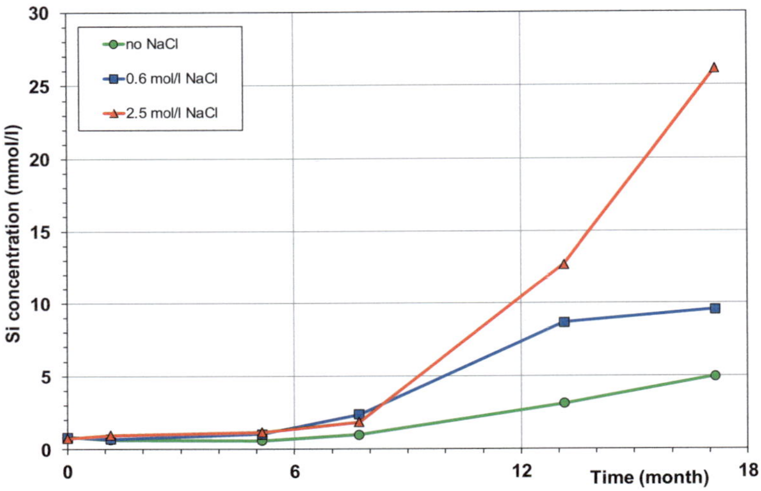

Fig. 2.7 Si concentration in a KOH solution (pH 13.9, $Ca(OH)_2$-saturated) with rhyolite grains (0–0.5 mm) and different NaCl concentration at 45 °C [70]

It was frequently questioned whether NaCl can penetrate sufficiently deep into concrete to drive an ASR. Often, only a low NaCl penetration of around 10–30 mm was found. This was pursued in a recent study by means of laser-induced breakdown spectroscopy (LIBS) on concrete prisms exposed to NaCl solution during the climate simulation concrete prism test (CS-CPT) as well as on untreated reference prisms [74]. It was found that the penetration depths of Na and Cl in the prisms exposed to the NaCl solution would have been underestimated (around 30–40 mm), if not compared to untreated reference prisms. The comparison to the Na and Cl profiles for the untreated prisms showed much higher depths of penetration (>80 mm). For field structures, undisturbed reference samples are usually not available for determining the original concentrations of Na and Cl. Therefore, it is likely that earlier penetration depths of Na and Cl on field samples have been underestimated.

2.2.5.2 The Influence of Alkali Acetates and Formates

There are several reports about severe ASR damage to airfield concrete pavements (e.g., [116, 138]). For example, a 12-year-old runway in Germany showed severe ASR damage 5–7 years after the de-icer was changed in the early 1990s from synthetic urea to potassium acetate [71]. In the following years more airfields seemed to be affected (Fig. 2.8) but no official data exist that could provide the total number of airports suffering from ASR and the extent of the damage. After all, a more recent report states that "many" airports in Germany are affected by ASR [1].

Unlike NaCl, no studies about the influence of alkali acetates and formates on ASR were available before 2005. First studies showed that concrete prisms exposed

Fig. 2.8 Airfield in Germany (2013, image by Colin Giebson) with severe map cracking and heaved precast drainage elements due to the expansion of the slabs

to de-icer solutions based on alkali acetates and especially alkali formates showed much higher expansions compared with reference prisms exposed to water or NaCl solution (Fig. 2.9). Further investigations showed an increase of the pH in solutions of the alkali acetates and formates after $Ca(OH)_2$ was added [55, 159]. The reason for this effect was found in [71, 73, 159] with the increase of the $Ca(OH)_2$ solubility due to the formation of Ca-acetate and Ca-formate complexes in such solutions (Fig. 2.10):

$$CH_3COO^- + K^+ + Ca^{2+} + 2OH^- \leftrightarrow CaCH_3COO^+(aq) + K^+ + 2OH^-$$

$$2\,CH_3COO^- + 2\,K^+ + Ca^{2+} + 2OH^- \leftrightarrow Ca(CH_3COO)_2(aq) + 2K^+ + 2OH^-$$

$$HCOO^- + K^+ + Ca^{2+} + 2OH^- \leftrightarrow CaHCOO^+(aq) + K^+ + 2OH^-$$

$$2\,HCOO^- + 2K^+ + Ca^{2+} + 2OH^- \leftrightarrow Ca(HCOO)_2(aq) + 2K^+ + 2OH^-$$

$Ca(OH)_2 \leftrightarrow Ca^{2+} + 2OH^-$ (further dissolution of $Ca(OH)_2$ after the depletion of Ca^{2+} by the formation of the Ca-acetate/formate complexes)

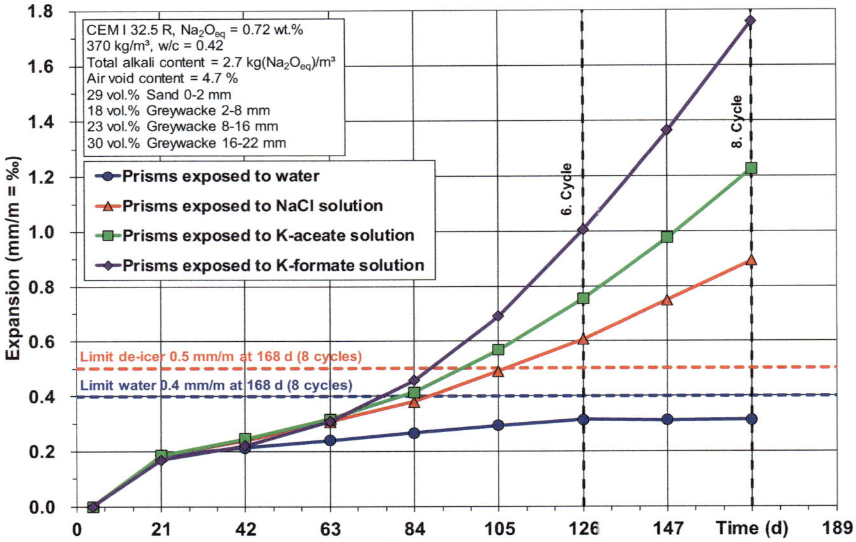

Fig. 2.9 Influence of different de-icer solutions on expansion during the ASR performance test (CS-CPT) (modified from [71])

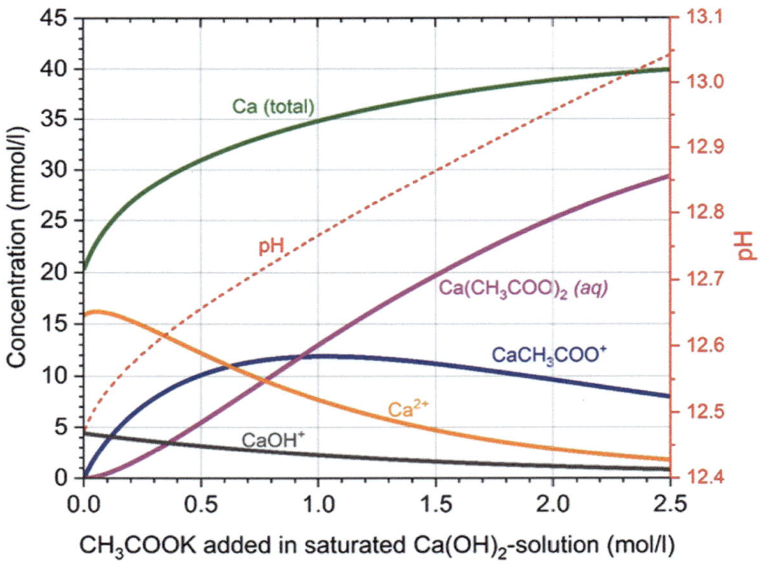

Fig. 2.10 Species distribution and pH in a saturated solution of $Ca(OH)_2$ with addition of CH_3COOK (based on [71, 73, 158]

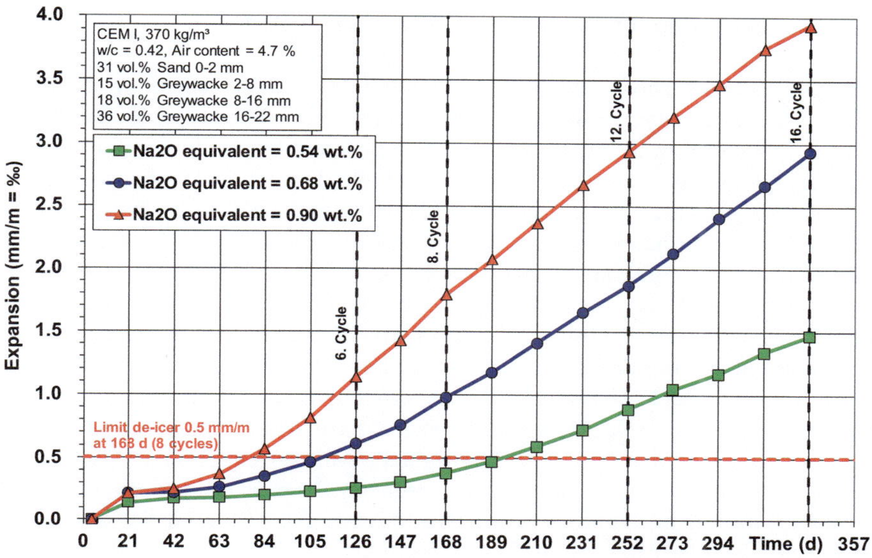

Fig. 2.11 Expansion during the ASR performance test (CS-CPT) for three airfield pavement concretes with reactive greywacke and different cement Na_2O_{eq} exposed to K-acetate de-icer solution [73]

Ultimately, the excess supply of alkalis and the release of OH^- ions due to the increased solubility of $Ca(OH)_2$ in the concrete pore solution can trigger or accelerate ASR in concrete that contains reactive aggregates and is exposed to de-icers based on alkali acetates and formates. This mechanism promotes the attack of reactive aggregates and the formation of ASR gel. The increase of the pH explains the more severe effect of the alkali acetates and formates compared to NaCl. Because of this special mechanism, low-alkali cements may only delay the ASR but cannot prevent it permanently (Fig. 2.11).

Another result of the study in Giebson [71] was an increase of the SO_4^{2-} concentration in pore solutions after cement paste samples were exposed to potassium acetate and especially potassium formate solutions. The increase of the SO_4^{2-} concentration was companied by a decomposition of ettringite, leading to an (unknown) S-free ettringite analogue and K_2SO_4. Another study found a severe leaching-dissolution effect on C-S-H paste samples immersed in a Ca-Mg-acetate solution [57]. Degradation effects on cement paste samples exposed to acetate solutions were also reported in further studies [104, 142]. These findings suggest effects besides ASR that are not yet sufficiently well investigated.

2.2.5.3 Conclusions on the Influence of De-Icing in ASR

Many studies have shown that alkali-containing de-icers can trigger or accelerate ASR in concrete with reactive aggregates. Depending on the type of the de-icer, ASR is influenced in different ways and to a different extent. In case of NaCl as the most commonly used de-icer in the world, the dissolution of silica in reactive aggregates is increased beyond the effect known from the pH-driven dissolution. In case of the alkali acetates/formates, mainly used for airfields, the pH in the concrete pore solution and consequently the dissolution of silica in reactive aggregates is greatly increased. Hence, together with the delivered alkalis, an ASR can be highly promoted in concrete exposed to NaCl or alkali acetate/formate-based de-icers. Experience has shown that the impact of the de-icers clearly grows in the sequence NaCl–Na/K-acetate–Na/K-formate. There is first evidence that former penetration depths of the de-icers (NaCl) into field concrete have been underestimated because reference samples for determining the original alkali concentration are usually not available anymore. However, all measures reducing the penetration depth of the de-icers are helpful in preventing a de-icer driven ASR. The ASR potential of concrete compositions under exposure of de-icer solutions can be tested with the CS-CPT (climate simulation concrete prism test), explained in detail elsewhere (e.g., [71, 73, 154]). This test is especially designed for pavement concretes and since 2018 part of the German guideline for building highway pavements [67].

2.3 Electrochemical Treatments

Electrochemical treatment techniques are methods which prevent the onset of reinforcement corrosion in concrete or decrease the rate to a tolerable level. The most common electrochemical technique is cathodic protection; however, other techniques such as realkalization, chloride extraction and electro-osmotic methods can be used to mitigate the risk of ongoing corrosion in reinforced concrete structures. Cathodic protection can also be optimized as a cathodic prevention system for installation during construction.

On the principle that corrosion occurs due to the formation of anodic sites where metal is lost (a) and cathodic sites which are unaffected (b), electrochemical techniques work by polarizing the reinforcement in an electrical circuit making all the steel to be protected cathodic with respect to a system of installed anodes, see Fig. 2.12.

$$\text{Anodic reaction}: \text{Fe} \rightarrow \text{Fe}^{2+} + 2\text{e}^-$$

$$\text{Cathodic reaction}: \text{O}_2 + 2\text{H}_2\text{O} + 4\text{e}^- \rightarrow 4\text{OH}^-$$

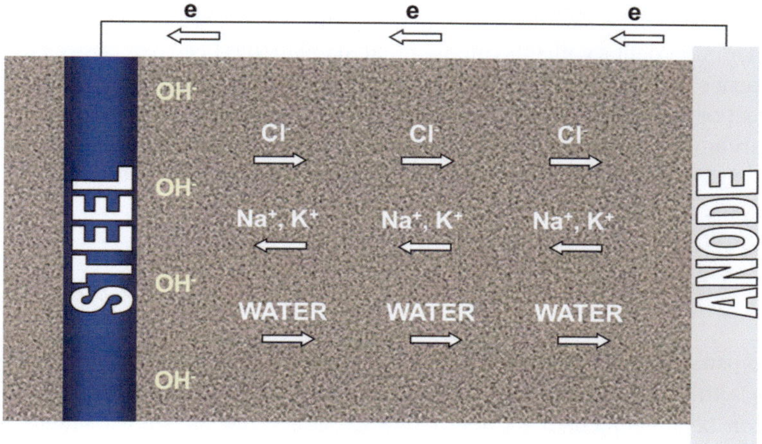

Fig. 2.12 Schematic of cathodic protection/electrochemical techniques (prepared by Rene Brueckner)

The electrons released at the anode react with oxygen and water to generate hydroxyl ions at the cathode due to electrolysis. In addition, because the cathode is negatively charged, ions such as chloride are repelled from the steel and attracted to the anodes via electro-osmosis. Cations, such as sodium and potassium, are however attracted to the cathode. Water moves towards the anode. Migration of ions between cathode and anode are considered to be beneficial side effects in the electrochemical rehabilitation of structures and used in optimised cathodic protection systems such as electro-osmosis, realkalization and chloride extraction.

2.3.1 *Cathodic Protection*

Cathodic protection systems can be self-powered galvanic or, more commonly, based on inert anodes powered by a low voltage DC supply, which is commonly referred to as impressed current cathodic protection (ICCP). The latter system requires an impressed current density of approximately 2–20 mA/m^2 [5] which generates hydroxyl ions at the cathode and facilitates the migration of chloride, sodium and potassium ions within the cover zone of the concrete over the lifetime of the system. The current density depends on the degree of chloride contamination and exposure conditions.

The alkalinity at the cathode will increase due to the generation of hydroxyl ions and the accumulation of sodium and potassium ions from the concrete cover zone and external sources such as de-icing salts penetrating the cover zone. The migration of alkalis from the bulk of concrete towards the reinforcement is limited and only the reinforcement near zone experiences ion migration. The driving force as a result of the applied current density is relatively low. ISO 12696 [5] states that cathodic protection applied in accordance with this International Standard has been demonstrated to have no influence on ASR. However, literature presents some contradictory findings about the effect of cathodic protection in ASR. According to some authors [147, 169] the application of this electrochemical technique increases the risk of ASR, while in others (e.g., [168]) there are no deleterious effects of cathodic protection.

2.3.2 Electro-Osmosis

Electro-osmosis is the process where an impressed electric current is able to move liquids within pores of a porous material towards the negative electrode. This principle can be implemented on a structure to reduce the moisture content using a controlled low voltage DC pulse. Such systems are capable of reducing moisture levels in concrete to between 60 and 70% RH, and maintaining this level irrespective of external weather conditions [103].

Electro-osmosis has an additional benefit through the removal of excess free moisture. The level of dissolved salts present within the pore solution of the concrete, notably chlorides, can be reduced to below critical threshold for corrosion to occur. The reduction of the moisture content also reduces the risk of potential alkali-silica reactions despite the increase of alkalinity at the reinforcement.

2.3.3 Realkalization

While the beneficial secondary processes occur naturally as a consequence of cathodic protection, they can also be optimised as remediation treatments. Those optimised techniques can be applied temporarily in order to increase the alkalinity or reduce chloride levels by using much higher current densities and voltages as short-term treatments.

The technique of electrochemical realkalization (EcR) restores the alkaline environment around the reinforcement when lost as a result of carbonation. The target pH value is greater than 10.5 which is sufficient to restore and maintain the protective oxide layer on the reinforcement. Typically, in the realkalization process, a temporary mesh electrode contained within a reservoir of alkaline sodium or potassium carbonate solutions is attached to the surface of the concrete. An electrical connection is made between the surface mesh and the reinforcement, and a current is applied, such that the reinforcement becomes cathodic and the mesh anodic. This is the same principle as cathodic protection, but the applied current density is in the region of 1 A/m^2 of steel surface area but may be increased up to 2 and 4 A/m^2 in accordance with BS EN 14038–1 [64] and NACE SP0107 [126], respectively. The voltage is in the range of 30 to 50 V direct current. This means that the driving force of moving alkalis towards the reinforcement is significantly greater.

The alkaline sodium or potassium carbonate ions from the reservoir migrate to the cathode via electro-osmosis. Electrolysis at the cathode simultaneously results in the generation of hydroxyl ions which additionally increase the alkalinity around the reinforcement and so restores the passive layer on the steel surface. When the cover zone is saturated (confirmed by testing) the current is disconnected, and the external electrode and reservoir are removed. A beneficial side effect of the realkalization process is the repellence of chloride ions from the cathode. The chloride content can be significantly reduced despite a smaller total current being applied than with electrochemical chloride extraction.

ASR susceptible structures may be at risk from this type of remediation method. If it is known that the structure is potentially susceptible to ASR, then it is possible to vary the treatment by using lithium borate solutions so that repassivation solely relies on the electrolysis and the risk is significantly reduced.

2.3.4 Electrochemical Chloride Extraction

Electrochemical chloride extraction (ECE), sometimes erroneously referred to as desalination, makes use of the same electrochemical principles as realkalization, except that the primary aim is to remove negatively charged chloride ions from within the concrete by ion migration to a surface mounted inert anode.

The current density required for chloride extraction to be effective is generally much greater than that required for realkalization but should not exceed a maximum of 4 and 10 A/m^2 of steel surface area in accordance with NACE SP0107 [126] and CEN-TS-14038–2 [3]. Therefore, electrolysis at the reinforcement and the migration of alkalis is significantly increased. The electrolyte is required to be above pH 6 to prevent the formation of chlorine gas. This can be achieved by using potable water as electrolyte that is however to be regularly replaced to maintain the minimum pH of 6. Alkaline electrolytes minimise the risk of acidification of the electrolyte and the evolution of chlorine gas. Saturated calcium hydroxide with an excess of solid material to maintain the saturation is a suitable alkaline solution but NACE SP0107 [126] recommend that lithium borate solution should be used where there is a concern of ASR.

The time period required to reduce the chloride concentration at the reinforcement to acceptable levels is dependent on a number of factors such as cement type, density, concrete cover and ambient temperature but is generally between 6 to 8 weeks (e.g., [143]).

References

1. Anonymous (2016): Betonkrebs am Flughafen? Internes Papier warnt vor maroden Pisten. http://www.n-tv.de/wirtschaft/Internes-Papier-warnt-vor-maroden-Pisten-article17442081.html. Beitrag vom 12.04.2016, Accessed 23 Feb 25
2. BS EN 1008: Mixing water for concrete. Specification for Sampling, Testing and Assessing the Suitability of Water, Including Water Recovered from Processes in the Concrete Industry, as Mixing Water for Concrete. British Standards Institution, 22 pp (2002)
3. CEN-TS-14038–2: Electrochemical Re-alkalization and Chloride Extraction Treatments for Reinforced Concrete—Part 2: Chloride extraction. CEN, 14 pp (2011)
4. Agrolanzarote: Hojas informativas, Servicio insular agrario [online] (2012). Accessed 2018
5. ISO 12696: Cathodic protection of steel in concrete. International Organization for Standardization, Geneva, Switzerland, 45 pp (2012)
6. AASHTO M 144: Standard specification for calcium chloride. American Association of State Highway and Transportation Officials (AASHTO), 5 pp (2014)
7. AASHTO M 154: Standard specification for air-entraining admixtures for concrete. American Association of State Highway and Transportation Officials (AASHTO), 2 pp (2011)
8. AASHTO M 194M/M 194: Standard specification for chemical admixtures for concrete. American Association of State Highway and Transportation Officials (AASHTO), 11 pp (2013)
9. ACI 212.3R-10: Report on chemical admixtures for concrete. Reported by ACI Committee 212, Farmington Hills, American Concrete Institute, USA, 7 pp (2010)
10. Aitcin, P.C., Flatt, R.J.: Science and Technology of Concrete Admixtures. Woodhead Publishing, 666 pp (2016)
11. ASTM D98: Standard specification for calcium chloride. ASTM International, West Conshohocken, USA, 3 pp (2005)
12. ASTM C1603: Standard test method for measurement of solids in water. ASTM International, West Conshohocken, USA, 4 pp (2016)
13. ASTM C260: Standard specification for air-entraining admixtures for concrete. ASTM International, West Conshohocken, USA, 4 pp (2016)
14. ASTM C114: Standard test methods for chemical analysis of hydraulic cement. ASTM International, West Conshohocken, USA, 32 pp (2018)
15. ASTM C1017/C1017M: Standard specification for chemical admixtures for use in producing flowing concrete. ASTM International, West Conshohocken, USA (Withdrawn 2022) (2013)
16. ASTM C1602/C1602M: Standard specification for mixing water used in the production of hydraulic cement concrete. ASTM International, West Conshohocken, USA, 5 pp (2012)
17. ASTM C494/C494M: Standard specification for chemical admixtures for concrete. ASTM International, West Conshohocken, USA, 10 pp (2017)
18. ASTM C94/C94M: Standard specification for ready-mixed concrete. ASTM International, West Conshohocken, USA, 146 pp (2017)
19. ASTM C125: Standard terminology relating to concrete and concrete aggregates. ASTM International, West Conshohocken, USA, 9 pp (2017)
20. ASTM C162: Standard terminology of glass and glass products. ASTM International, West Conshohocken, USA, 16 pp (2015)
21. ASTM C289: Standard test method for potential alkali-silica reactivity of aggregates (Chemical method). ASTM International, West Conshohocken, USA, 7 pp (2007)
22. Bayliss, P., Kolitsch, U., Nickel, E.H., Pring, A.: Alunite supergroup: recommended nomenclature. Mineral. Mag. **74**(5), 919–927 (2010)
23. Bender, T., Karagülle, Z., Bálint, G.P., Gutenbrunner, C., Bálint, P.V., Sukenik, S.: Hydrotherapy, balneotherapy, and spa treatment in pain management. Rheumatol. Int. **25**(3), 220–224 (2005)
24. Berry, E.E., Malhotra, V.M.: Fly ash in concrete. Department of Energy, Mines and Resources, Canada Centre for Mineral and Energy Technology, 178 p (1986)

25. Bérubé, M.A., Fournier, B.: Alkalis releasable by aggregates in concrete: significance and test methods. In: Tang, M., Deng, M. (eds.) Proceedings of the 12th International Congress on Alkali-Aggregate Reactions in Concrete (ICAAR), Beijing, China, pp. 17–30 (2004)
26. Bérubé, M.A., Frenette, J.: Testing concrete for AAR in NaOH and NaCl solutions at 38 °C and 80 °C. Cement Concr. Compos. **16**(3), 189–198 (1994)
27. Bérubé, M.A., Dorion, J.F., Vézina, D.: Labouratory and field investigations of the influence of sodium chloride on alkali-silica reactivity. In: Bérubé, M.A., Fournier, B., Durand, B. (eds.) Proceedings of the 11th International Conference on Alkali-Aggregate Reaction, Québec City, Canada, pp. 149–158 (2000)
28. Bérubé, M.A., Duchesne, J., Dorion, J.F., Rivest, M.: Labouratory assessment of alkali contribution by aggregates to concrete and application to concrete structures affected by alkali-silica reactivity. Cem. Concr. Res. **32**(8), 1215–1227 (2002)
29. Bhatty, J.I.: Alternative uses of cement kiln dust. Research and Development Information, Portland Cement Association, pp. 1–17 (1995)
30. Blum, A.E.: Feldspars in weathering. In: Parsons, I. (ed.) Feldspars and their reactions, NATO ASI Series, C421, pp. 595–630. Kluwer Academic Publishers, Dordrecht (1994)
31. Boles, J.R., Johnson, K.S.: Influence of mica surfaces on pore-water pH. Chem. Geol. **32**, 303–317 (1984)
32. Bond, R.G., Straub, C.P.: CRC Handbook of environmental control, vol. III, pp. 671–686. Water Supply and Treatment.CRC Press, Boca Raton, FL, USA (1980)
33. Bonnet, P.: Gonflement du béton de plusieurs barrages. Proc. Journées d'étude sur les réactions liants-granulats dans les bétons. École Nationale des Ponts et Chaussées, Paris, France, 54 pp (1989)
34. Broekmans, M.A.T.M.: Deleterious reactions of aggregate with alkalis in concrete. Rev. Mineral. Geochem. **74**(1), 279–364 (2012)
35. Brown, P., Bothe, J.: The system $CaO–Al_2O_3–CaCl2–H_2O$ at 23±2 °C and the mechanisms of chloride binding in concrete. Cem. Concr. Res. **34**(9), 1549–1553 (2004)
36. Brueckner, R.: Schematic of cathodic protection/electrochemical techniques, personal communication (2014)
37. BS 3148: Methods of Test for Water for Making Concrete (Including Notes on the Suitability of the Water). British Standards Institution, 10 pp (1980)
38. UC: Propiedades del agua de mar. Geografía del mar (Properties of seawater. Geography of sea) [online], Pontificia Universidad Católica de Chile, Santiago de Chile (2018). Accessed 2022
39. Chao, S.H., Hargreaves, A., Taylor, W.H.: The structure of orthoclase. Mineral. Mag. J. Mineral. Soc. **25**(168), 498–512 (1939)
40. Chatterji, S.: Freezing of air-entrained cement-based materials and specific actions of air-entraining agents. Cement Concr. Compos. **25**(7), 759–765 (2003)
41. Chatterji, S., Jensen, A.D., Thaulow, N., Christensen, P.: Studies of alkali-silica reaction. Part 3. Mechanisms by which NaCl and $Ca(OH)_2$ affect the reaction. Cem. Concr. Res. **16**(2), 246–254 (1986)
42. Chatterji, S., Thaulow, N., Jensen, A.D.: Studies of alkali-silica reaction. Part 4. Effect of different alkali salt solutions on expansion. Cem. Concr. Res. **17**(5), 777–783 (1987)
43. Chester, R., Jickells, T.D.: Marine Geochemistry, 3th ed. Wiley-Blackwell (2012)
44. Choquette, M., Bérubé, M.-A., Locat, J.: Behaviour of common rock-forming minerals in a strongly basic NaOH solution. Can. Mineral. **29**, 163–173 (1991)
45. Chou, L., Wollast, R.: Steady-state kinetics and dissolution mechanisms of albite. Am. J. Sci. **285**(10), 963–993 (1985)
46. Coombs, D.S., Alberti, A., Armbruster, T., Artioli, G., Colella, C., Galli, E., Grice, J.D., Liebau, F., Mandarino, J.A., Minato, H., Nickel, E.H., Passaglia, E., Peacor, D.R., Quartieri, S., Rinaldi, R., Ross, M., Sheppard, R.A., Tillmanns, E., Vezzalini, G.: Recommended nomenclature for zeolite minerals: report of the subcommittee on zeolites of the International Mineralogical Association, Commission on New Minerals and Mineral Names. Can. Mineral. **35**, 1571–1606 (1997)

47. Crundwell, F.K.: The mechanism of dissolution of minerals in acidic and alkaline solutions: Part I—a new theory of non-oxidation dissolution. Hydrometallurgy **149**, 252–264 (2014)
48. Crundwell, F.K.: The mechanism of dissolution of minerals in acidic and alkaline solutions: Part II—Application of a new theory to silicates, aluminosilicates and quartz. Hydrometallurgy **149**, 265–275 (2014)
49. Davis, G., Oberholster, R.E.: The alkali-silica reaction product. A mineralogical and an electron microscopic study. Proceedings of the Eighth International Conference on Cement Microscopy, Florida, USA (1986)
50. De Weerdt, K., Justnes, H.: The effect of seawater on the phase assemblage of hydrated cement paste. Cement Concr. Compos. **55**, 215–222 (2015)
51. De Weerdt, K., Orsáková, D., Müller, A.C.A., Larsen, C.K., Pedersen, B., Geiker, M.R.: Towards the understanding of chloride profiles in marine exposed concrete, impact of leaching and moisture content. Constr. Build. Mater. **120**(1), 418–431 (2016)
52. De Weerdt, K., Lothenbach, B., Geiker, M.R.: Comparing chloride ingress from seawater and NaCl solution in Portland cement mortar. Cem. Concr. Res. **115**, 80–89 (2019)
53. Deer, W.A., Howie, R.A., Wise, W.S., Zussman, J.: Rock Forming Minerals, Volume 4B, Framework Silicates: Silica Minerals Feldspathoids and the Zeolites, 2nd edn. Geological Society of London, UK, 982 pp (2004)
54. Demayo, A.: Elements in seawater. In: Lide, D.R. (eds.) CRC Handbook of Chemistry and Physics. CRC Press, U.S.A., pp. 14–10 (1992)
55. Diamond, S., Kotwica, L., Olek, J., Rangaraju, P.R., Lovell, J.: Chemical aspects of severe ASR induced by potassium acetate airfield pavement deicer solution. In: Proceedings of the 8th CANMET International Conference on Recent Advances in Concrete Technology, Marc-André Bérubé Symposium on Alkali-Aggregate Reactivity in Concrete, Montreal, Canada, pp. 261–277 (2006)
56. Dove, P.M., Elston, S.F.: Dissolution kinetics of quartz in sodium chloride solutions: Analysis of existing data and a rate model for 25 °C. Geochim. Cosmochim. Acta **56**, 4147–4156 (1992)
57. Dramé, H., Beaudoin, J.J., Raki, L.: A comparative study of the volume stability of C-S-H (I) and Portland cement paste in aqueous salt solutions. J. Mater. Sci. **42**, 6837–6846 (2007)
58. Dressler, A.: Einfluss von Tausalz und puzzolanischen, aluminiumhaltigen Zusatzstoffen auf die Mechanismen einer schädigenden Alkali-Kieselsäure-Reaktion in Beton. Dissertation, Technische Universität München(2013)
59. Dressler, A., Urbonas, L., Heinz, D.: Effect of alkali ingress and binder composition on damaging ASR. In: Drimalas, T., Ideker, J.H., Fournier, B. (eds.) Proceedings of the 14th ICAAR, Austin, Texas, USA, 10 pp (2012)
60. Drolet, C., Duchesne, J., Fournier, B.: Effect of alkali release by aggregates on alkali-silica reaction. Constr. Build. Mater. **157**, 263–276 (2017)
61. Duchesne, J., Bérubé, M.A.: Evaluation of the validity of the pore solution expression method from hardened cement pastes and mortars. Cem. Concr. Res. **24**(3), 456–462 (1994)
62. Duchesne, J., Bérubé, M.A.: Effect of supplementary cementing materials on the composition of cement hydration products. Adv. Cem. Based Mater. **2**(2), 43–52 (1995)
63. Durth, W., Hanke, H.: Handbuch Straßenwinterdienst. Kirschbaum Verlag, Bonn (2004)
64. BS EN 14038–1: Electrochemical Realkalization and Chloride Extraction Treatments for Reinforced Concrete—Part 1: Realkalization. British Standards Institution, 22 pp (2016)
65. Etxeberria, M., Corominas, A.G., Pardo, P.: Influence of seawater and blast furnace cement employment on recycled aggregate concretes' properties. Constr. Build. Mater. **115**(15), 496–505 (2016)
66. Ferraz, A.R., Fernandes, I., Soares, D., Santos Silva, A., Quinta-Ferreira, M.: Influence of the alteration of granitic rocks on alkalis release. World Multidisciplinary Earth Science Symposium—WMESS 2017, Prague, 9 pp (2017)
67. FGSV: TP B-StB—Teil 1.1.10, AKR-Potenzial und Dauerhaftigkeit von Beton (Klimawechsellagerung), Ausgabe (2018)
68. Fibrometals: Fibras metálicas [online] (2018). Available from https://www.fibrometals.ro/en-us/Fibers. Accessed 2025

69. Gartiser, S., Reuther, R., Gensch, C.O.: Machbarkeitsstudie zur Formulierung von Anforderungen für ein neues Umweltzeichen für Enteisungsmittel für Straßen und Wege, in Anlehnung an DIN EN ISO 14024. Umweltbundesamt, Berlin (2003)
70. Giebson, C.: Lösevorgänge im Gesteinskorn und Quellverhalten von AKR-Gelen. Vortrag zum 4. Workshop der DFG-FOR 1498 „AKR unter kombinierten Einwirkungen", Karlsruher Institut für Technologie (KIT), Karlsruhe (2017)
71. Giebson, C.: Die Alkali-kieselsäure-reaktion in Beton für Fahrbahndecken und Flugbetriebsflächen unter Einwirkung alkalihaltiger Enteisungsmittel. Dissertation, Bauhaus-Universität Weimar (2013). http://nbn-resolving.de/urn:nbn:de:gbv:wim2-20131217-20916
72. Giebson, C., Seyfarth, K., Ludwig, H.M.: Influence of sodium chloride on ASR in highway pavement concrete. In: Bernardes, H., Hasparyk, N.P. (eds.) Proceedings of the 15th ICAAR, São Paulo, Brazil, 10 pp (2016)
73. Giebson, C., Seyfarth, K., Stark, J.: Influence of acetate and formate-based deicers on ASR in airfield concrete pavements. Cem. Concr. Res. **40**, 537–545 (2010)
74. Giebson, C., Voland, K., Ludwig, H.M., Meng, B.: Alkali-silica reaction performance testing of concrete considering external alkalis and preexisting microcracks. Struct. Concr. **18**, 528–538 (2017)
75. Gillot, J.E., Rogers, C.A.: Alkali-aggregate reaction and internal release of alkalis. Mag. Concr. Res. **46**, 99–112 (1994)
76. Grattan-Bellew, P.E., Beaudoin, J.J.: Effect of phlogopite mica on alkali-aggregate expansion in concrete. Cem. Concr. Res. **10**, 789–797 (1980)
77. Grattan-Bellew, P.E., Danay, A.: Comparison of labouratory and field evaluation of AAR in large dams. In: Proceedings of the International Conference on Conc AAR in Hydroelectric Plants and Dams, Canadian Electrical Association (CEA), Montreal, in association with Can Natl ICOLD Committee, Fredericton, Canada, pp. 1–23. Hansen, W.C. (1959). Release of alkalies by sands and admixtures in portland cement mortars, ASTM Bull., pp. 35–38 (1992)
78. Hanna, J., Dorota, S.L., Alicja, G.: Application in concrete plants waste recovered from concrete industry. In: International RILEM Conference on the Use of Recycle Materials in Buildings and Structures, Barcelona, Spain, pp. 1–9 (2004)
79. Hartman, D., Greenwood, M.E., Miller, D.M.: High Strength Glass Fibers. AGY Technical Paper (1996)
80. Hawthorne, F.C., Oberti, R., Harlow, G.E., Maresch, W.V., Martin, R.F., Schumacher, J.C., Welch, M.: IMA report—nomenclature of the amphibole supergroup. Am. Miner. **97**(11–12), 2031–2048 (2012)
81. Herholdt, A.D., Justesen, C.F.P., Nepper-Christensen, P., Nielsen, A.: Beton bogen, Aalborg Portland, 354 pp (1979)
82. Hooton, R.D., Julio-Betancourt, G.A.: Chemical attack of various chloride salts on concrete. In: Stark, J (ed.) Proceedings of the 17th ibausil, Weimar, Germany, pp. 13–23 (2009)
83. Hooton, R.D., Thomas, M.D.A., Ramlochan, T.: Use of pore solution analysis in design for concrete durability. Adv. Cem. Res. **22**(4), 203–210 (2010)
84. INDURA: Neutralización de aguas alcalinas con CO_2 (Neutralization of alkaline water with CO_2) [online] (2018). Available from http://www.indura.cl/Web/CL/Menu/206. Accessed 2018
85. Infoagro Systems: The fertilisers [online] (2018). Available from http://www.infoagro.com/abonos/abonos_y_fertilizantes2.htm. Accessed 2022
86. IQB: Agua alcalina (Alkaline water) [online] (2017). Available from http://www.iqb.es/cbasicas/farma/farma06/agua/agua_alcalina.htm. Accessed 2022
87. Jakobsen, U.H., De Weerdt, K., Geiker, M.R.: Elemental zonation in marine concrete. Cem. Concr. Res. **85**, 12–27 (2016)
88. Jawed, I., Skalny, J.: Alkalies in cement: a review I. Forms of alkalies and their effect on clinker formation. Cem. Concr. Res. **7**(6), 719–730 (1977)
89. Ji, D., Lou, Z., He, M., Shi, Y., Gu, X.: Effect of both grafting and blending modifications on the performance of lignosulphonate-modified sulphanilic acid–phenol–formaldehyde condensates. Cem. Concr. Res. **42**(9), 1199–1206 (2012)

90. Juel, I.: Mineralogical and Thermodynamic Processes by Sulphate and Seawater Attack of Danish Concrete, in Faculty of Science. University of Copenhagen, Geological Institute (2002)
91. Juenger, M., Provis, J.L., Elsen, J., Matthes, W., Hooton, R.D., Duchesne, J., Courard, L., He, H., Michel, F., Snellings, R., De Belie, N.: Supplementary Cementitious Materials for Concrete: Characterization Needs. Mater. Res. Soc. Symp. Proc. Vol. 1488, Materials Research Society (2012)
92. Katayama, T.: ASR gels in concrete subject to freeze-thaw cycles—Comparison between labouratory and field concretes from Newfoundland, Canada. In: Proceedings of the of the 13th ICAAR, Trondheim, Norway, pp. 174–183 (2008)
93. Katayama, T.: ASR gels and their crystalline phases in concrete—universal products in alkali-silica, alkali-silicate and alkali-carbonate reactions. In: Proceedings of the 14th ICAAR, Austin, Texas, USA, paper 030411-DTA-03, 12 pp (2012)
94. Kawamura, M., Ichise, M.: Characteristics of alkali-silica reaction in the presence of sodium and calcium chloride. Cem. Concr. Res. **20**, 757–766 (1990)
95. Kawamura, M., Koike, M., Nakano, K.: Release of alkalis from reactive andesite aggregates and fly ashes into pore solution in mortars. In: Okada, K, Nishibayashi, S., Kawamura, M. (eds.) Proceedings of the 8th International Conference on Alkali-Aggregate Reaction, Kyoto, Japan, pp. 271–278 (1989)
96. Kawamura, M., Takemoto, K., Terashima, N.: Effect of sodium chloride and sodium hydroxide from the surrounding solution on alkali-silica reaction in mortars containing fly ash. Mag. Concr. Res. **40**, 143–151 (1988)
97. Kawamura, M., Takeuchi, K., Sugiyama, A.: Mechanisms of expansion of mortars containing reactive aggregate in NaCl solution. Cem. Concr. Res. **24**, 621–632 (1994)
98. Kawamura, M., Takeuchi, K., Sugiyama, A.: Mechanisms of the influence of externally supplied NaCl on the expansion of mortars containing reactive aggregate. Mag. Concr. Res. **48**, 237–248 (1996)
99. Ketcham, S.A., Minsk, L.D., Blackburn, R.R., Fleege, E.J.: Manual of practice for an effective anti-icing program—A guide for highway winter maintenance personnel. U.S. Department of Transportation, Federal Highway Administration, Publication No. FHWA-RD-95-202 (1996)
100. Kim, B.G., Jiang, S.J., Jolicoeur, C., Aïtcin, P.C.: The adsorption behaviour of PNS superplasticizer and its relation to fluidity of cement paste. Cem. Concr. Res. **30**(6), 887–893 (2000)
101. Kosmatka, S., Kerkhoff, B., Panerese, W.: Design and control of concrete mixtures (14 ed.). Portland Cement Association, Skokie, Illinois (2002)
102. Kucche, K.J., Jamkar, S.S., Sadgir, P.A.: Quality of water for making concrete: a review of literature. Int. J. Sci. Res. Publ. **5**(1), 1–10 (2015)
103. Lambert, P.: Controlling moisture, construction repair. Concrete Repairs 6 (1997)
104. Lee, H., Cody, R.D., Cody, A.M., Spry, P.G.: Effects of various deicing chemicals on pavement concrete deterioration. Mid-Continent Transportation Symposium, Center for Transportation Research and Education, Iowa State University, pp. 151–155 (2000)
105. Leemann, A., Lothenbach, B., Thalmann, C.: Influence of superplasticizers on pore solution composition and on expansion of concrete due to alkali-silica reaction. Constr. Build. Mater. **25**(1), 344–350 (2011)
106. Lindgård, J., Çakır, O.A., Fernandes, I., Rønning, T.F., Thomas, M.D.A.: Alkali–silica reactions (ASR): Literature review on parameters influencing labouratory performance testing. Cem. Concr. Res. **42**(2), 223–243 (2012)
107. Locati, F., Marfil, S., Baldo, E., Maize, P.: Na_2O, K_2O, SiO_2 and Al_2O_3 release from potassic and calcic-sodic feldspars into alkaline solutions. Cem. Concr. Res. **40**, 1189–1196 (2010)
108. Locher, F.W., Sprung, S.: Ursache und wirkungsweise der alkalireaktion. In: Walz, K. (Herausgeber): Betontechnische Berichte 1973. Beton-Verlag GmbH Düsseldorf, S. pp. 101–123 (1974)
109. Low, G.L., Ng, K., Ng, W.L., Tam, C.T., Heng, R.B.W.: Use of recycled cement-based slurry water for making concrete. J. Inst. Eng. Malaysia **68**(4), 47–55 (2007)

110. Lu, D., Zhou, X., Xu, Z., Lan, X., Tang, M., Fournier, B.: Evaluation of labouratory test method for determining the potential alkali contribution from aggregate and the ASR safety of the Three-Gorges dam concrete. Cem. Concr. Res. **36**, 1157–1165 (2006)
111. Maas, A.J., Ideker, J.I., Juenger, M.C.G.: Alkali silica reactivity of agglomerated silica fume. Cem. Concr. Res. **37**(2), 166–174 (2007)
112. Malmstrøm, M., Banwart, S.: Biotite dissolution at 25 °C: The pH dependence of dissolution rate and stoichiometry. Geochim. Cosmochim. Acta **61**, 2779–2799 (1997)
113. Maltese, C., Pistolesi, C., Bravo, A., Cella, F., Cerulli, T., Salvioni, D.: A case history: effect of moisture on the setting behaviour of a Portland cement reacting with an alkali-free accelerator. Cem. Concr. Res. **37**(6), 856–865 (2007)
114. MARASAS: Labouratorio y servicios agropecuarios (n.d.) Composición química de fertilizantes (Chemical composition of fertilizers) [online] (2018). Available from https://laboratoriomarasas.ar/pdfs/COMPOSICION-QUIMICA-DE-FERTILIZANTES_0.pdf. Accessed 2025
115. Marshall, W.L., Warakomski, J.M.: Amorphous silica solubilities II. Effect of aqueous salt solutions at 25°C. Geochim. Cosmochim. Acta **44**, 915–924 (1980)
116. Martin, L.C.: Alkali-aggregate reaction in portland cement concrete (PCC) airfield pavements, Engineering Technical Letter (ETL) 06-2, U.S. Air Force Civil Engineer Support Agency (2006)
117. McCoy, W.J., Eshenour, O.L.: Significance of total and water-soluble alkali contents of Portland cement. J. Mater. **3**(3), 684–694 (1968)
118. Mehta, P.K.: Studies on chemical resistance of low water/cement ratio concretes. Cem. Concr. Res. **15**(6), 969–978 (1985)
119. Mehta, P.K., Monteiro, P.J.M.: Concrete: Microstructure, Properties, and Materials, 3rd edn., p. 659. McGraw-Hill, New York, USA (2013)
120. Menéndez, E.: Analysis of Atlantic seawater, IETcc-CSIC. Institute of construction sciences Eduardo Torroja belonging to the Spanish National Research Council (2017)
121. Menéndez, E.: Estudio microstructural de productos de reacción álcali-árido en hormigones curados a alta temperatura (Microstructural study of alkali-aggregate reaction products in concretes curing to high temperature). Materiales de construcción, Vol. 43, nº 232, oct-dic (1993)
122. Menéndez, E., Prendes, N., Márquez, C., Aldea, B.: Evaluations of granitic aggregates behaviour in relation with the alkaline extraction and compositional change in their phases. In: Proceedings of 14th International Conference on Alkali-Aggregate Reactions in Concrete, Austin-Texas (USA), 10 pp (2012)
123. Mizumoto, Y., Kosa, K., Ono, K., Nakano, K.: Study of cracking damage of a concrete structure due to alkali-silica reaction. In: Proceedings of the 7th International Conference on AAR, Concrete Alkali-aggregate Reaction, Noyes Publications, New York, pp. 204–209 (1987)
124. Morimoto, N., Fabries, J., Ferguson, A.K., Ginzburg, I.V., Ross, M., Seifert, F.A., Zussman, J., Aoki, K., Gottardi, G.: Nomenclature of pyroxenes. Am. Miner. **73**, 1123–1133 (1988)
125. Myrdal, R.: Accelerating admixtures for concrete. SINTEF REPORT Building and Infrastructure Concrete and Innovation Center, Project No: 3D006020 (2007)
126. NACE SP0107: Electrochemical realkalization and chloride extraction for reinforced concrete. NACE International, Houston, USA, 22 pp (2017)
127. Nagy, K.L.: Dissolution and precipitation kinetics of sheet silicates, In: White, A.F., Brantley, S.L. (eds.) Chemical weathering rates of silicate minerals. Reviews in Mineralogy, vol. 31, pp.173–233 (1995)
128. Nixon, P.J., Canham, I., Page, C.L., Bollinghaus, R.: Sodium chloride and alkali—aggregate reaction. In: Proceedings of the 7th International Conference on Alkali–Aggregate Reactions in Concrete, Ottawa, Canada, pp. 110–114 (1987)
129. Nixon, P.J., Page, C.L., Canham, I., Bollinghaus, R.: Influence of sodium chloride on the ASR. Adv. Cem. Res. **1**, 99–105 (1988)

130. Rieder, M., Cavazzini, G., D'yakonov, Y.S., Frank-Kamenetskii, V.A., Gottardi, G., Guggenheim, S., Koval, P.V., Müller, G., Neiva, A.M.R., Radoslovich, E.W., Robert, J.L., Sassi, F.P., Takeda, H., Weiss, Z., Wones, D.R.: Nomenclature of the micas. Can. Mineral. **36**, 905–912 (1998)
131. Oelkers, E.H., Schott, J., Gauthier, J.-M., Herrero-Roncal, T.: An experimental study of the dissolution mechanism and rates of muscovite. Geochim. Cosmochim. Acta **72**, 4948–4961 (2008)
132. U.S. ONR: Ocean water: temperature [online] (2007). Available from https://web.archive.org/web/20071212151229/http://www.onr.navy.mil/Focus/ocean/water/temp3.htm. Accessed 2022
133. Ouali, A.: Étude expérimentale du lessivage des alcalins et du silicium extractibles des granulats naturels, p. 176. Université de Paris VI, France (1997)
134. Ouyang, X., Qiu, X., Chen, P.: Physicochemical characterization of calcium lignosulfonate—a potential useful water reducer. Colloids Surf. A **282–283**(20), 489–497 (2006)
135. Parmelee, C.W., Monack, A.J.: The solubility of feldspars in water. J. Am. Ceram. Soc. **13**, 386 (1930)
136. Ramachandran, V.S.: Concrete Admixtures Handbook, 2nd ed., Properties, Science and Technology. William Andrew Publishing, 1183 pp (1996)
137. Ramezanianpour, A.A.: Cement Replacement Materials, p. 336. Springer-Verlag, Berlin Heidelberg, Springer Geochemistry/Mineralogy (2014)
138. Rangaraju, P.R., Olek, J.: Potential for Acceleration of ASR in the Presence of Pavement Deicing Chemicals. IPRF Research Report (2007)
139. Ribbe, P.H.: Chemistry, structure and nomenclature of feldspars. In: Ribbe, P.H. (eds.) Feldspar Mineralogy, 2nd edn. Reviews in Mineralogy, 2, pp. 1–19 (1983)
140. Rixom, R., Mailvaganam, N.: Chemical Admixtures for Concrete, 3rd edn., p. 456. E&FN Spon, London (1999)
141. Salvador, R.P., Cavalaro, S.H.P., Segura, I., Figueiredo, A.D., Pérez, J.: Early age hydration of cement pastes with alkaline and alkali-free accelerators for sprayed concrete. Constr. Build. Mater. **111**(15), 386–398 (2016)
142. Santagata, M.C., Collepardi, M.: The effect of CMA deicers on concrete properties. Cem. Concr. Res. **30**, 1389–1394 (2000)
143. Sergi, G., Page, C.L., Thompson, D.M.: Electrochemical induction of ASR in concrete. Mater. Struct. **24**, 359–361 (1991)
144. Seyfarth, K., Giebson, C.: Beurteilung des AKR-Schädigungspotentials von Betonen mittels Klimawechsellagerung. Beton- und Stahlbetonbau (100), Tagungsband Doktorandensymposium 45. Forschungskolloquium des DAfStb., Wien, Österreich, S., pp. 189–192 (2005)
145. Seyfarth, K., Giebson, C., Stark, J.: Prevention of deleterious ASR by assessing aggregates and specific concrete mixtures. In: Tasnimi, A.A. (ed.) Proceedings of the 3rd International Conference on Concrete and Development, Tehran, Iran, pp. 159–169 (2009)
146. Shayan, A.: Effects of NaOH and NaCl solutions and temperature on the behaviour of specimens subjected to accelerated AAR tests. Cem. Concr. Res. **28**, 25–31 (1998)
147. Shayan, A., Xu, A., Pritchard, R.: Influence of cathodic protection impresse currents on AAR expansion of concrete containing reactive aggregates. In: Proceedings of the 14th International Conference on Alkali Aggregate Reaction in Concrete, Austin, Texas, USA, 10 pp (2012)
148. Shayan, A., Xu, A., Chirgwin, G., Morris, H.: Effects of seawater on AAR expansion of concrete. Cem. Concr. Res. **40**(4), 563–568 (2010)
149. Shehata, M.H., Thomas, M.D.: Alkali release characteristics of blended cements. Cem. Concr. Res. **36**(6), 1166–1175 (2006)
150. Sibbick, R.G., Page, C.L.: Effects of sodium chloride on the alkali-silica reaction in hardened concretes. In: Shayan, S. (ed.), Proceedings of the 10th ICAAR, Melbourne, Australia, pp. 822–829 (1996)
151. Sibbick, R.G., Page, C.L.: Mechanisms affecting the development of alkali-silica reaction in hardened concretes exposed to saline environments. Mag. Concr. Res. **50**, 147–159 (1998)

152. Sibbick, T., Fenn, D., Crammond, N.: The occurrence of thaumasite as a product of seawater attack. Cement Concr. Compos. **25**(8), 1059–1066 (2003)
153. Siddique, R., Khan, M.I.: Supplementary Cementing Materials, Engineering Materials. Springer-Verlag Berlin Heidelberg 2011, XVI, 288 p (2011)
154. Sims, I., Poole, A.B. (eds.): Alkali-Aggregate Reaction in Concrete: A World Review. CRC Press/Balkema, Taylor & Francis Group (2017)
155. CSA Standard A23.2–27A: Standard practice to identify potential for alkali-reactivity of aggregates and measures to avoid deleterious expansion in concrete. Canadian Standards Association, Mississauga, Ontario, Canada (2014)
156. Stark, D.: Alkali-silica reactivity—some considerations. Cem. Concr. Aggregates **2**(2), 92–94 (1980)
157. Stark, D., Bhatty, M.S.Y.: Alkali-silica Reactivity: Effect of Alkali in Aggregate on Expansion. American Society for Testing and Materials, pp. 16–30 (1986)
158. Stark, J., Giebson, C.: Assessing the durability of concrete regarding ASR. In: Malhotra, V.M. (ed.) Proceedings of the 7th CANMET/ACI International Conference on Durability of Concrete, Montreal, Canada, pp. 225–238 (2006)
159. Stark, J., Giebson, C.: Influence of acetate and formate based deicers on ASR in airfield concrete pavements. In: Broekmans, M.A.T.M., Wigum, B.J. (eds.) Proceedings of the 13th ICAAR, Trondheim, Norway, pp. 665–674 (2008)
160. Stark, J., Seyfarth, K.: Performance testing method for durability of concrete using climate simulation. In: Malhotra, V.M. (ed.), Proceedings of the 7th CANMET/ACI International Conference on Durability of Concrete, Montreal, Canada, pp. 305–326 (2006)
161. Stoica, G., Perez-Ramirez, J.: Stability and inter-conversion of synthetic dawsonites in aqueous media. Geochim. Cosmochim. Acta **74**, 7048–7058 (2010)
162. Tanaka, M., Takahashi, K.: Characterization of salting-out effect for silicate complexes in sodium chloride and nitrate solutions using a FAB-MS. Anal. Sci. **17**(Supplement), i21–i23 (2001)
163. Tanaka, M., Takahashi, K.: Study on the salting-out effect using silica species by FAB-MS. J. Solution Chem. **36**, 27–37 (2007)
164. Taylor, H.F.W.: Cement Chemistry, 2d edn., p. 459. Thomas Telford, London (1997)
165. Neville, A.M., Brooks, J.J.: Concrete Technology, 2nd ed. Pearson, Essex, England, 442 pp (2010)
166. Thaulow, N., Geiker, M.: Determination of the residual reactivity of alkali-silica reaction in concrete. In: Proceedings of the 9th International Conference on Alkali-Aggregate Reaction in Concrete, London, UK, pp. 1050–1058 (1992)
167. Thomas, M.: The effect of supplementary cementing materials on alkali-silica reaction: a review. Cem. Concr. Res. **41**(12), 1224–1231 (2011)
168. Torii, K., Prasetia, I., Minato, T., Ishii, K.: The feature of cracking in prestressed concrete bridge girders deteriorated by alkali-silica reaction. Proceedings of the 14th International Conference on Alkali Aggregate Reaction in Concrete, Austin, Texas, USA, 10 p (2012)
169. Torii, K., Kawamura, M., Matsumoto, K. Ishii, K.: Influence of cathodic protection on cracking and expansion of the beams due to alkali silica reaction. In: Shayan, S. (ed.) Proceedings of the 10th International Conference on Alkali Aggregate Reaction in Concrete, Melbourne, Australia, pp. 653–660 (1996)
170. Torre, F.: Características y usos de los aceros (Characteristics and use of steels) [online] (2018). Available from https://sites.google.com/site/conocerlosmateriales/home/proceso-de-fabricacion-los-aceros/caracteristicas-y-usos-de-los-aceros. Accessed 2022
171. Ucha, F.: Fibras sintéticas (Synthetic Fibers) [online]. Accessed 2025
172. Uchikawa, H., Hanehara, S., Sawaki, D.: The role of steric repulsive force in the dispersion of cement particles in fresh paste prepared with organic admixtures. Cem. Concr. Res. **27**(1), 37–50 (1997)
173. UNE 83957:2008: Durabilidad del hormigón. Aguas de amasado y aguas agresivas. Determinación del residuo seco (Durability of concrete. Aggressive and mixing waters), UNE Normalización (2008)

174. Urd Sylte, G.: Den aller kaldaste havstraumen, Forskning [online] (2010). Available from https://forskning.no/havforskning-klima-vaer-og-vind/2010/05/den-aller-kaldaste-hav straumen. Accessed 2025
175. Van Aardt, J.H.P., Visser, S.: Calcium hydroxide attack on feldspars and clays: possible relevance to Cement-aggregate reactions. Cem. Concr. Res. **7**(6), 643–648 (1977)
176. Vernon, R.H.: A Practical Guide to Rock Microstructure, p. 606. Cambridge Univ. Press, United Kingdom (2004)
177. Visvesvaraya, H.C., Mullick, A.K., Samuel, G., Sinha, S.K., Wason, R.C.: Alkali reactivity of granitic rock aggregates. In: Proceedings of the 8th Congress on Chemical Cements, pp. 206–213 (1986)
178. Wang, Y., Yu, G., Deng, M., Tang, M., Lu, D.: The use of thermodynamic analysis in assessing alkali contribution by alkaline minerals in concrete. Cem. Concr. Compos. **30**(4), 353–359 (2008)
179. Wang, Y., Deng, M., Tang, M.: Alkali release from aggregate and the effect on AAR expansion. Mater. Struct. **41**, 159–171 (2008)
180. Whitney, D.L., Evans, B.W.: Abbreviations for names of rock-forming minerals. Am. Miner. **95**, 185–187 (2010)
181. Wille, W., Hettche, H., Clodius, C.D.: Enteisungsabwasser von flughäfen. Arbeitsgemeinschaft Deutscher Verkehrsflughäfen (ADV), Stuttgart (1998)
182. Woermann, E., Hahn, T., Eysel, W.: The substitution of alkalies in tricalcium silicate. Cem. Concr. Res. **9**, 701–711 (1979)
183. Wollast, R.: Kinetics of the alteration of K-feldspar in buffered solutions at low temperature. Geochimica Cosmochimica Acta **31**, 635–648 (1967)
184. Xu, Q., Stark, J.: Early hydration of ordinary Portland cement with an alkaline accelerator. Adv. Cem. Res. **17**(1), 1–8 (2005)
185. Yamada, K., Kawabata, Y., Igarashi, G., Sagawa, Y.: Importance of the "pH of concrete" in various aspects of concrete durability and performance. In: Proceedings of the 15th International Congress on the Chemistry of Cement, Prague, Check Republic (2019)
186. Yazici, H.: The effect of silica fume and high-volume class C fly ash on mechanical properties, chloride penetration and freeze-thaw resistance of self-compacting concrete. Constr. Build. Mater. **22**(4), 456–462 (2008)
187. Zhang, S., Liu, Y.: Molecular-level mechanisms of quartz dissolution under neutral and alkaline conditions in the presence of electrolytes. Geochem. J. **48**, 189–205 (2014)

Chapter 3
Analytical Methods to Determine the Alkalis

Maarten A. T. M. Broekmans, Esperanza Menéndez Méndez, and Eric Giannini

Abstract This chapter discusses methods of analysis of concrete and aggregate samples to determine their alkali content and potential availability to contribute to alkali-silica reaction (ASR). Sample size and representativity are important considerations for any test method, and these are discussed as a pre-requisite to sample acquisition and preparation. Preparation methodologies appropriate to aggregates and concrete are discussed, with detailed guidance offered for a range of analytical techniques. Guidance on conducting the analysis by wet chemical methods, X-ray fluorescence (XRF), atomic absorption spectroscopy (AAS), inductively coupled plasma spectroscopy (ICP), in-situ elemental analysis by scanning electron microscopy (SEM) or electron microprobe analysis (EMPA), and ion-selective electrodes is presented. Finally, guidance is provided on reporting analytical results and accounting for analytical error.

Keywords Alkalis · Concrete · Aggregates · Alkali-aggregate reaction · Alkali-silica reaction · Chemical analysis · SEM-EDS · EMPA · XRF · AAS · ICP · Sample preparation · Specimen preparation

---

Esperanza Menéndez Méndez—Deceased 21 November 2024.

This chapter was written with the cooperation of Cedric Drolet, Josée Duchesne and Ricardo García-Roves.

M. A. T. M. Broekmans (✉)
Geological Survey of Norway-NGU, Department of Mineral Resources/Department of Laboratories, P. O. Box 6315 Torgården, 7491 Trondheim, Norway
e-mail: maarten.broekmans@ngu.no

E. Menéndez Méndez (Deceased)
Institute Eduardo Torroja of Construction Science (IETcc-CSIC), C/ Serrano Galvache, 4, 28033 Madrid, Spain
e-mail: emm@ietcc.csic.es

E. Giannini
Portland Cement Association, 200 Massachusetts Ave NW, Suite 200, Washington, DC 20001, USA
e-mail: egiannini@cement.org

E. Menéndez Méndez and A. Santos Silva (eds.), *Total Alkalis in Concrete—Contribution to Alkali-silica Reaction*, RILEM State-of-the-Art Reports 40, https://doi.org/10.1007/978-3-031-56164-1_3

3.1 Sample Size and Representativity

3.1.1 General Considerations

Despite it is not often enough acknowledged as such, 'sample preparation' *s.l.* represents an essential part of any analytical method. A poor sampling strategy as well as poor sample (or specimen) preparation strictly reduce analytical repeatability, inhibit reproduction of obtained of results, and may thus render an otherwise suitable analytical method unreliable. Analytical protocol as well as instrumentation are designed for robustness and will diligently produce a measurement result, no matter the nature of the sample material they are fed, neither the treatment the material received to transform it from 'sample as collected' into 'specimen prepared for analysis'.

Obviously, even the most rigorous sample preparation protocol and/or most precise analytical method are not able to transform a non-representative sample into a representative one. Thus, sample representativity must be esteemed as an integrated part of sample acquisition and preparation, whether of massive rock (e.g., natural rock or anthropogenic rock - hardened concrete), of particulate solids (e.g., recycled concrete, virgin aggregate, cement, SCMs), fluids and liquids (e.g., mortar mix additives), or other.

> **Definition**: a sample is considered representative if the error caused by the sampling protocol, is subordinate to error caused by the entire analytical protocol, including sample and specimen preparation as inalienable parts thereof.

The above definition means that for example, the iron content in multiple samples subsequently extracted from one and the same iron-ore stockpile source, all are within analytical error from one another. In other words, the variation attributable to the sampling process between individual samples from the same 'parent' stays within analytical error range. Thus, any variation greater than the analytical error is 'genuine' (as opposed to 'sampling artefact') and must hence reflect real variation in the sample material.

While a given sample may be representative for one method of assessment, it may not be for another analytical method that is able to assess the same sample property/quality with greater level of accuracy. Alternatively, a given sample may be representative for a certain predefined 'species of interest' (e.g., lithologies - sandstone, mylonite; chemical constituents - Na, K), but not so for other species (e.g., chert/flint, or Li), depending on their relative abundance (in vol%, wt%, or other). Thus, the analytical method of choice and minimum representative sample size are mutually interdependent.

If in a particulate material the individual particles constitute the 'counting unit' (i.e., as in particle counting practice), then a sample's representativity is constrained by the total number of particles it contains. This is entirely governed by mathematics of 'particle count statistics', which determines the minimum error in vol%. The mathematical background was initially devised by Gy [100–102] for mineral processing in the ore industry. Over the years, Gy's insights have become known as the 'Theory of Sampling' - ToS, which has found global acceptance in the mining industry, and has been extended to comprise also concrete, aggregate, and other mineral materials (e.g., [104, 117–120, 145, 190, 227, 243]).

The mathematical background demonstrates that 'sample representativity' is a composite parameter uniting a number of independent variables that are not always easy to grasp, each contributing but not always in a straight-forward manner, and at times even counter-intuitive. Therefore, the calculated minimum sample size required to be genuinely representative cf. above definition, typically exceeds so-called 'guesstimates' by a large factor. Mathematical statistics are rarely appreciated and commonly met with skepticism and ignorance (see e.g., [27, 122]), but yet and despite do provide fundamental understanding for how to conduct reliable analysis [241]. Contrary to popular belief, the mathematical science of statistics is not defined by 'extensive experience', neither can it be 'outsmarted' [122]. An initially non-representative sample cannot be made (more) representative by after treatment of any kind ([177], pp. 23).

As an example, for a species of interest - SOI with relative abundance within 15–85 vol%, assessment of at least one thousand (1000) individual particles is required to obtain a 10%-relative error. Thus, when all 1000 particles present in the as-received sample are counted, the true abundance of a constituent measured at 20.0 vol% in reality lies in the range 18–22 vol%, i.e., 20.0 ± 2.00 vol%. To achieve the same error level for a constituent with relative abundance <15 vol%, more than one thousand particles need to be counted overall, to compensate for the fact that too few 'particles of interest' (for abundance >85 vol%, too few 'particles *not* of interest') are present in a 1000-particle population to keep error within 10%-relative.

To illustrate this, consider a SOI at 15.0 vol% abundance in a 1000-particle sample, which is thus represented by one hundred and fifty individual particles. One particle more or less makes a difference of one in one hundred and fifty, which has only a marginal effect on the total number counted (151 or 149 particles), i.e., only 0.1 vol% on 15.0 vol%, or ~1/150th, which equals ~0.67%-relative. In contrast, a SOI at 5 vol% abundance in a 100-particle sample is represented by only 5 particles. One particle less or more changes it the species' abundance to 4 vol% or 6 vol%, or one in five. That represents *a relative* change of 20%, i.e., 30× times larger than in the previous example. To achieve the same 0.67%-relative error as for the 150-in-1000 example, the 5 vol%-abundance sample would need to contain at least 3000 particles, for the 5% SOI to be represented by 150 particles. Thus, it is the abundance of the SOI combined together with the desired relative error that determine overall sample size.

Table 3.1 provides relative errors for the petrographic assessment of fine aggregate <4 mm, assessed with either four, two, or even only one single standard-sized thin-section (~27 mm × 47 mm), separated into three narrower (alternative 1) or two broader (alternative 2) size fractions. The tabulated values demonstrate how relative error increases towards lower SOI abundances and for fewer particles counted, the more so for the broader size fractions in alternative 2.

The mathematical-statistical underpinning assumes all particles are of uniform size, so that the individual contribution of any given particle is identical to any other. If the particle size range in a material sample varies more than a factor 2–3×, then

Table 3.1 Examples of relative errors after assessment of 4, 2 or 1 thin-section(s) for aggregate constituents at five abundance levels, grouped per size fraction

	Total number assessed			Relative error (%) at SOI abundance level				
	Thin-sections	Particles	Counts	50 vol%	20 vol%	10 vol%	5 vol%	2 vol%
Alternative 1	Size fraction 2–4mm, ~150 particles/~250 counts per section, 2.00 mm grid							
	4	600	1000	8	16	24	32	48
	2	300	500	12	22	32	44	63
	1	150	250	16	32	44	64	80
	Size fraction 1–2mm, ~ 600 particles/~900 counts per section, 0.75 mm grid							
	4	2400	3600	4	8	12	16	24
	2	1200	1800	6	12	16	24	36
	1	600	900	8	16	24	32	48
	Size fraction 0.063–1mm, ~1800 particles/~2500 counts per section, 0.50 mm grid							
	4	7200	10000	*<2	*4	*6	*8	*12
	2	3600	5000	>3	7	10	15	22
	1	1800	2500	5	10	15	20	30
Alternative 2	Size fraction 2–4mm, ~150 particles/~250 counts per section, 2.00 mm grid							
	4	600	1000	8	16	24	32	48
	2	300	500	12	22	32	44	63
	1	150	250	16	32	44	64	80
	Size fraction 0.063–2mm, ~1200 particles/~1600 counts per section, 0.50 mm grid							
	4	4800	6400	<3	6	9	12	20
	2	2400	3200	4	8	12	16	24
	1	1200	1600	6	12	16	24	36

*: extrapolated from Figs. B.1 and B.2, Appendix B, RILEM AAR 1.1 (2016, p. 58)

the individual contribution of a small particle will 'weigh' substantially less than of a larger particle, as particle volume is a third-power function of particle (nominal) dimension. A doubling of nominal size (whether diameter for spherical particles, or edge for cubic) results in a 2^3 = eightfold increase in particle volume, and thus its weight (physically as well as statistically in particle counting). Thus, a field sample may be divided into a number of narrower size fractions by sieving, and each fraction assessed separately, to ensure reliable assessment [149, 150].

Alternatively, a wide particle size range can be uniformized by mechanical comminution ('crushing'). Comminution is often regarded as merely 'optional according to preference', but instead may represent a necessary preparation stage essential for reliable analysis of samples with a large particle size range (or solid samples with a large mineral grain size range). This has long been understood by the mining industry, especially for noble metals (e.g., gold, platinum metals) where sample representativity is quintessential for the economic survival of an extraction location.

3.1.2 Acquiring a Representative Sample

Sample representativity for particulate materials is elaborated in detail in RILEM AAR-1.1 [199] for petrographic assessment of coarse virgin aggregate materials, with special reference to minimum required sample sizes in kilograms for a number of (nominal) particle sizes from 4 to 50 mm (Table 3.1). Assuming a material density of 2650 kg·m^{-3} (as for granite or arkose) and uniform particle size, the stated values provide a sample containing minimum one thousand particles. If the density of a given aggregate material is greater than 2650 kg·m^{-3}, then the minimum sample weight should be increased proportionally. The statistical uncertainty for counting one thousand individual particles of equal statistical weight (i.e., uniform particle size) can be read from Figs. B.1 and B.2 in Appendix B of RILEM AAR-1.1 ([199], pp. 58). With one thousand particles, a 10%-relative error is achieved only for SOI with an abundance of 20 vol% or above. For any SOI with abundance <20 vol%, relative error is larger than 10%-relative, reaching 100%-relative for SOI with 1.0 vol% abundance.

The AAR-1.1 [199] guideline text does acknowledge the effect of low relative abundances of a given species of interest, by recommending additional minimum sample sizes. Figure 3.1 also contains a simple ^{2}log-curve (dashed black), showing how sample weight develops with increasing nominal particle size (2 mm–4 mm–8 mm–16 mm–32 mm–64 mm; truncated at 50 mm by the plot area), starting at the low end with the weight of one thousand cubes of 2 mm on edge of granitic/arkosic density. With uniform 2 mm diameter spheres instead of cubes, the same sample weight would contain many more particles, about 1900. However, this provides additional in sample representativity and further enhances robustness of the whole analytical procedure. Despite, remarkably, sample weights for materials with relative SOI abundances at 20 vol% (blue) and 10 vol% (green) are in fact *lower* than the weights given in default Table 1 (the red curve). In addition, Fig. 3.1 reveals that the default red curve only exceeds the ^{2}log-curve for particle sizes finer than ~22 mm (which in principle is good), whereas for coarser particles it dips under the ^{2}log-curve (which is not good). This means that sample weights according to the blue or green curves contain fewer than one thousand (1000) individual particles, as well as the red curve for particles coarser than >22 mm. Only the curve intended for SOIs with abundance at 5 vol% (orange) exceeds the simple ^{2}log-curve for one thousand granitic/arkosic particles.

Furthermore, it is important to notice that quite a few European countries have current national standards that suggest limiting chert/flint in aggregate to 3.0 vol% (UK), 2.0 vol% (NL), or even 1.0 vol% (DK). Due to ongoing research on field structures made with locally available materials, such national guidance is liable to change and limit contents may be adjusted to reflect the most recent experiences, usually towards lower values. Current Tables 1 and 2 in RILEM AAR-1.1 do not provide guidance for such low SOI abundances, and the numbers of particles needed for reliable analysis are multiple thousands instead, and consequently with much higher minimum representative sample weights (also see e.g., [27]).

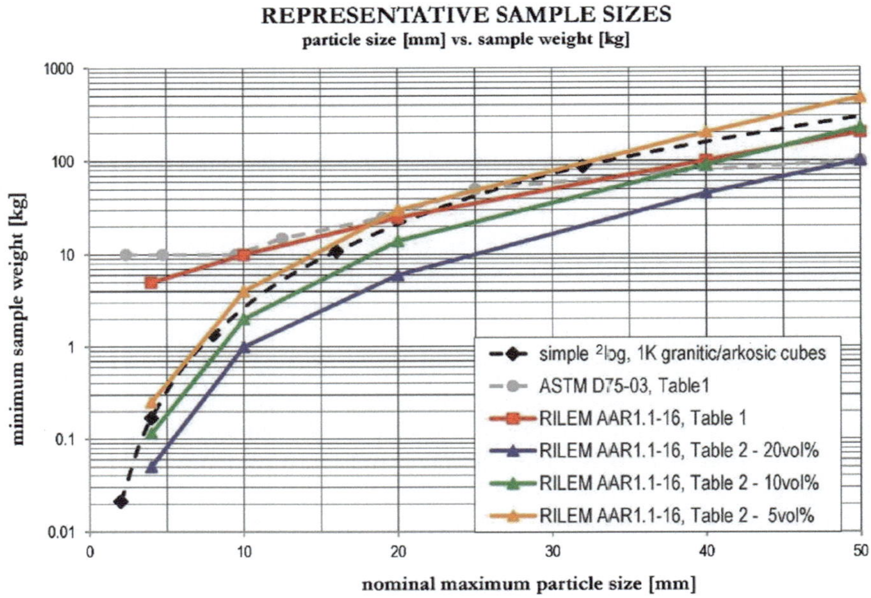

Fig. 3.1 Sample weights (kg) for one thousand particles of granitic density, according to RILEM AAR-1.1 [199]. Depending on the nature of the actual sample material under consideration, one thousand particles may or may not be representative. Further explanation in text

Table 3.2 provides minimum numbers of particles required in a sample, to be considered representative for a certain predefined analytical error level for a given SOI at a specific abundance, again mathematically determined by statistics. The minimum numbers of particles listed in the table are now used to calculate total sample weight using nominal particle size (mm) and lithological density (kg·m^{-3}) (see e.g., [31]).

In assessment daily practice, 10%-relative error is typically deemed acceptable. Table 3.2 convincingly demonstrates that even 50%-relative error (unacceptably high) requires assessment of a large number of particles for a SOI at 1.0 vol% abundance. Conversely, petrographic assessment of 1000 particles in a sample material containing a small amount of chert/flint (e.g., from UK, NL, DK, as mentioned above) has a 50%-relative uncertainty. For example, finding 2.0 ± 1.0 vol% chert/flint in a given aggregate material means its real content is somewhere between 2.0 − 1.0 = 1.0 vol% and 2.0 + 1.0 = 3.0 vol%, which may make the difference between acceptance of the material or its rejection, depending on the position of criteria (UK, NL, or DK, as mentioned above). In many cases, this kind of inconclusiveness may trigger reassessment, leading to additional cost and probably also time delay.

An experienced petrographic operator might in principle be capable of assessing a few thousand particles a day, but is of course subject to fatigue and tedium, which contribute to operator ennui and error, arguably increasing with the number particles

Table 3.2 Examples of *minimum* numbers of particles to be assessed, to achieve a predefined analytical error level for selected SOI abundance levels

SOI abundance (vol%)	Absolute error (vol%)	Net content range (vol%)	Minimum number of particles needed
Predefined error 50%-relative			
50	±25.0	25.0–75.0	~25
20	±10.0	10.0–30.0	50
10	±5.0	5.0–15.0	100
5	±2.5	2.5–7.5	200
2	±1.0	1.0–3.0	500
1	±0.5	0.5–1.5	1100
Predefined error 20%-relative			
50	±10.0	40.0–60.0	100
20	±4.0	16.0–24.0	350
10	±2.0	8.0–12.0	800
5	±1.0	4.0–6.0	1700
2	±0.4	1.6–2.4	4000
1	±0.2	0.8–1.2	*8800
Predefined error 10%-relative			
50	±5.0	45.0–55.0	400
20	±2.0	18.0–22.0	1500
10	±1.0	9.0–11.0	3200
5	±0.5	4.5–5.5	*7000
2	±0.2	1.8–2.2	*15000
1	±0.1	0.9–1.1	*33000

*: extrapolated from Figs. B.1 and B.2, Appendix B, RILEM AAR-1.1 [199, p. 58]

still waiting to be assessed. In addition, samples containing more than a few thousand particles become prohibitively expensive to assess because of hours spent on assessment. However, recent developments in SEM instrumentation and machine-learning software have now enabled automated identification of mineral constituents and rock types in particulate materials [105, 106]. This enables fully automated assessment of many multiple thousands of particles, nearly completely avoiding operator-related effects, shortening turn-around time, and with increased reliability [149].

An alternative mechanism causing concrete damage by expansion was recognized about two decades ago (e.g., [34]), and was more recently identified as the primary cause of damage in a housing project [201, 232]. The expansion is caused by as little as 0.1 wt% of total-sulfur ('total-S', present in very common sulfide minerals like e.g., pyrite (FeS_2), and especially pyrrhotite ($Fe_{1-x}S$). Currently, potentially reactive aggregate is tested for oxidation capacity in a specially designed reaction cell, as assessment by point-counting is not feasible when regarding the numbers of particles needed in a sample to be representative. Fully automated assessment by an instrumental method would greatly contribute to prevention of future cases.

3.2 Preparation of Sample Materials and Specimen for Analysis

3.2.1 Reducing Sample Size and Maintaining Representativity: Comminuting, Homogenizing, Splitting

The amount of sample material that is fed into the analytical instrument is only a small fraction of the bulk sample. For instance, a standard-sized petrographic thin-section of 28 mm × 48 mm = 1344 mm^2 at 30 μm = 0.03 mm thickness contains maximum 40.32 mm^3 of sample material, which volume equals 106.85 mg, assuming granitic or arkosic density (=2650 kg·m^{-3}). However, sample matter rarely ever covers thin-sections edge-to-edge for reason of preparation practicality, thinner sections contain proportionally less material, and the total volume assessed by SEM–EDS or EPMA is much, much smaller still by a large factor [188, 189].

For another example, in a typical flux-fusion procedure for X-Ray Fluorescence (XRF) bulk whole-rock analysis, 0.6 ± 0.0001 g (=600 mg) of pulverized sample material is digested in 4.2 ± 0.0001 g (= 7 × excess) Li-borate flux at ~1050 ± 25 °C to produce a glass bead, which is then analysed by the XRF spectrometer. The amount of powder used for digestion in excess strong acid for analysis by inductively-coupled plasma - atomic/optical emission spectrometry (ICP-AES/OES) is typically of the same size order, i.e., a few hundreds of milligrams.

Analytical results obtained on such tiny subsamples can only be correlated reliably to the bulk sample if representativity during reduction of an 'as received' sample to a subsample is strictly maintained. The definition of sample representativity given above applies unabridged to subsamples as well: variation between different subsamples from the same parent sample are to remain within analytical uncertainty.

Procedures and protocol to ensure this are elaborated in the published literature (e.g., [102, 103, 190, 192]), and comprise comminution, homogenization, and splitting. It is thus full well possible to analyse sub-gram portions sampled from a >5000-tonne train load of gold ore concentrate (72 ppm Au) at the destination (port), compare with the origin (quarry), and confirm that the material is identical and not diluted with material of lower grade [159]).

It must be reinstated that a non-representative 'as received' bulk sample cannot be made (more) representative, no matter how carefully it is processed or treated ([177], pp. 23).

3.2.1.1 Comminution

According to RILEM AAR-1.1 [199], a representative sample of coarse aggregate easily weighs well in excess of 10 kg, occasionally even considerably more. For analysis, the 'as received' sample must be reduced in size to suit the analytical method of choice (comprising both sample and specimen preparation), to produce a specimen that fits the relevant analytical instrument, e.g., microscope, spectrometer. The key issue during all steps of comminution, homogenizing and splitting is to maintain and preserve representativity of the 'as received' bulk sample, down to the final subsample (aka. test portion) that is ultimately analysed. That means that the few hundreds of milligrams contained in the prepared specimen for analysis (whether thin section, fusion bead, acid solution, or otherwise) need to be as statistically representative (in particle count) as the 'as received' bulk sample. Thus, the particle size of the final pulverized product for chemical assay is determined as the number of particles that need to be present in the final subsample. For example, a 600 mg test portion that needs to contain one million particles needs to be milled finer than a 600 mg test portion that only needs to contain one thousand particles. The number of particles needed to be representative for a given species of interest - SOI is statistically determined (as elaborated above), and can be very large for low-abundance species.

The first step in this is comminution, which is the generic term for particle or grain size reduction of solid materials by crushing, pulverizing, or diverse other methods. A broad choice of equipment and machinery is available from many manufacturers worldwide, from (cone, jaw, hammer) crushers with capacities sufficient to serve a large quarry, down to benchtop models capable of crushing samples of tennis ball size. Crushing equipment is commonly designed to handle sample material feed in a continuous process, which may be practically limited by particle size, but in principle not by overall sample tonnage. Among the most common types for laboratory use is the jaw crusher, which combines robust and straight-forward operation with effective crushing action and easy to clean design. The particle size of the final crushed product can be set within certain limits to meet demand, and especially for smaller mills a choice of jaw materials (e.g., various steel qualities, tungsten carbide, zirconia) is available to reduce wear and increase longevity.

It should be noted that a (laboratory) jaw crusher is normally designed to handle brittle sample materials of intermediate hardness and tenacity (i.e., 'rocks'), to produce a uniform crushed product.

Ideally, the comminuted product has a high cubicity (i.e., any particle's orthogonal axes x, y, z are of equal length, no elongate or flaky particles), as well as a narrow particle-size distribution. Sound, modern hardened concrete and virgin aggregate will normally fit these characteristics. However, the cement paste and/or aggregate in 'damaged concrete *s.l.*' may be altered (in older structures often of lower strength-class too), increasing hardness and tenacity contrast between aggregate and binder.

This causes severe intergrinding during crushing, when the harder particles contribute to the comminution of the softer, resulting in a distinctly bimodal particle size distribution of the crushed product, in such cases consisting of coarse hard particles (aggregate) and very fine dust (mostly cement paste). Also, natural polymictic gravel aggregate as well as any rock types with distinct mineral layering (e.g., gneiss, sandstone, limestone) with contrasting hard and soft lithologies are prone to inter-grinding. This kind of crushed product with a bimodal particle-size distribution is undesirable, as it hampers reliable splitting for preparation of subsamples, and requires additional comminution to again remove the attributed bimodality.

In the case of ASR-damaged concrete, this was done by crushing all bimodal bulk sample materials in a spring-loaded roller-press mill to crush coarse particles down to <0.5 mm, whereas finer particles simply fell through mostly unaffected. Reanalysis confirmed that this additional step successfully reduced bimodality and restored sample representativity (for full details, see [24]). Alternatively, the two fractions could have been separated by sieving, before additional comminution of the coarse fraction and recombining with the finer fraction to obtain a (more) unimodal distribution.

For certain sample materials, it may be possible (i.e., practicable) to separate different constituents by hand prior to crushing, i.e., when the number of constituents is limited, when they are well-liberated from one another, and for a limited overall number of particles. Inter-grinding can then be minimized by processing each constituent-fraction separately. Cleaning of equipment between successive (constituent) lithologies belonging to one single sample is not necessary in such cases.

For chemical assay, crushed material must first be pulverized to maximize its surface area so as to enhance digestion and/or dispersion. In contrast to crushing, however, laboratory pulverization equipment is normally designed for batch processing, in which a limited amount of material (typically up to about 50 g) is further comminuted to fine powder <20 μm. Therefore, the pre-crushed sample needs to be reduced in amount to fit the pulverization equipment, often in several iterations. To maintain representativity, rigorous procedures for homogenization and splitting are essential, which will be addressed in the next section.

Many different types of laboratory equipment for pulverization exist, e.g., mortar mills (i.e., motorized mortar and pestle), disc mills, and ball mills, available in many variants from many vendors. One popular option is the vibratory disc mill (aka. 'shatter box'), which combines effective milling action with adequate capacity and sufficiently fine final particle size, as well as it is relatively convenient to clean. The pre-crushed and split (sub-) sample material is charged manually into the milling vessel containing a ring and puck, which is then agitated vigorously by specially designed machine for a preset time.

As the contact between sample material and comminution equipment becomes increasingly intense with decreasing particle size, contamination with wear material from the mill itself becomes inevitable. To minimize the chance for contamination, vendors usually offer milling vessels with different material options for lining, puck and ring, e.g., diverse steel qualities, tungsten carbide, zirconia, toughened ceramics, and natural agate (see e.g., [113, 248], also see [98]). Latter lining option is very common in geological laboratories, as it is extremely hardwearing and attributes only minimal contamination to the sample material (e.g., [263]). Moreover, as the pale-grey natural agate as used for mill vessel lining is chemically composed of

>99.8 wt% SiO_2 (silica) and very little else, it will essentially only affect the generally most abundant constituent (silica, *in chemical terms!*), by an amount within analytical error at typical concentrations in siliceous supracrustal rocks. Carbonate rocks that contain much less intrinsic silica are mechanically much softer, which will wear off less material from the agate lining.

It should be emphasized that the pulverization equipment and procedures described in this text section are intended for analysis by one or another type of chemical assay. Thus, prime focus is on effective routine processing of many samples for digestion in excess solvent (whether HT-flux, acid, alkaline, or other), as opposed to 'preserving crystal-structural integrity' which is lost in digestion anyway. Therefore, for (crystal-structural) assessment by X-ray powder diffraction (XRPD), additional milling with different equipment is necessary, described elsewhere.

Crushing and milling equipment must be cleaned thoroughly before processing each next sample, taking apart all parts in contact with sample material and wiping with lint-free cloth or tissue and alcohol. Ultrasonication may be necessary to remove particles from inaccessible nooks or slots in removable parts. Non-removable parts and the machine base are carefully vacuum cleaned and wiped. Use of pressurized gas is strongly discouraged as it merely blows around sample particles in an uncontrollable manner contaminating the entire laboratory, and often leaves a greasy residue, making the next cleaning job even more arduous.

NB: airborne mineral dust is a possible health hazard!

Blunted sample material adhering to the crusher jaws can be removed with a hard steel brush, or abraded by brief milling with pure quartzite, which is inexpensive as well as effective for that purpose, while risk for contamination of siliceous supracrustal rocks is minimal. Truly good equipment design also facilitates convenient frequent cleaning.

3.2.1.2 Splitting and Homogenizing

The total weight of the crusher-product needs to be reduced to suit the pulverizing equipment, while maintaining overall representativity (as defined by particle-count statistics). Going from kilograms to grams comprises a thousand-fold reduction. As the representativity of the 'as received' particulate sample is governed by particle count statistics (see above), reduction to subsample size demands preservation of particle count statistics, which is conveniently achieved by particle size reduction through iterative comminution (i.e., crushing, grinding), with homogenization and splitting before each next step. Regardless of the procedure or machinery chosen, the key issue is to maintain (preserve) sample representativity, and to attribute minimal artefacts to the sample by the splitting.

The simplest way of splitting a sample of particulate material in two is by pouring it over a knife-edge. The knife can be made by sharply bending a rectangular plate of stainless steel into a reverse V, placed on a tray as above, in the middle under a chute or funnel through which the whole sample is loaded. One half of the sample is collected for further processing, the other half for storage and future use.

The next simplest way of subdividing sample material equipment-wise is by 'coning and quartering'. The crushed product is poured to form a right circular conical heap on a clean tray to prevent loss of any sample material (alternatively, a large clean sheet of kraft paper). If the cone base ('directrix') meets the tray (paper) edges, then a larger tray is needed. Next, the cone is flattened vertically into a cake using appropriate tools (e.g., shovel, trowel) avoiding any circular motion, then halved and each portion again halved, thus obtaining four sample portions of equal size. Now, two diagonally opposite quarters of the cone are combined for further processing, and the remaining material (indicated as 'reject') stored for future use. The quartering and recombining opposite quadrants (so, in effect, *halving*) intends to somewhat reduce sampling error by averaging out two and two portions.

A slightly more sophisticated way of splitting uses a riffle box (colloquial: 'riffler'). The bulk sample is loaded into a rectangular holder on the top of the equipment, spreading it evenly over the holder. The sample is then poured by some sort of tipping mechanism over an array of equally sized and spaced chutes (usually from six and up in a single riffle box) that divide the sample material over two pans at the bottom. Smaller riffle models box do without such holder, and the sample material is poured or scooped directly onto the chutes array. Riffle boxes are available in a variety of materials, capacities in L, chute sizes and slot numbers, to suit a range of overall sample sizes and particle sizes.

Above procedures may be reiterated to obtain a suitably sized subsample (test portion). As an example, consider a representative sample of 25 mm aggregate, cf. ASTM D75 [2] weighing 50 kg. The whole 50 kg have been comminuted in a jaw crusher to <2 mm, now, a subsample needs to be prepared for subsequent pulverization. The maximum capacity of the vibratory disc mill is ~45 g (a rather typical value for laboratory equipment), so the subsample needs to be on the order of 35–45 g. According to Fig. 3.1, a subsample of that weight will contain ~1000 particles <2 mm (cf. the ^{2}log curve, lower left corner). If ~1000 particles are adequate to achieve the desired level of uncertainty for the defined SOI (species of interest) in the bulk material, then initial crushing to <2 mm is adequate. However, if the predefined acceptable analytical error demands a greater number of particles, then the bulk sample needs to be crushed to a finer particle size to meet this requirement. It is advisable to ensure that the minimum number of particles needed is well exceeded, to avoid possible scarcity (especially if additional analysis should be required).

Reducing 50 kg (=50000 g) of crushed material down to 35–45 g thus requires ten (10) iterations (viz. 50000–25000–12000–6000–3000–1500–750-375-168-84-42); a much smaller sample weighing only 2.5 kg still requires six (6) iterations (2500-1250–625-313-157-78-39). This makes the coning and quartering method laborious and error prone to operator (in-) experience and inter-operator variation, most notably as multiple iterations on a single sample material will escalate an initially small incremental error. In particular, the method is sensitive for bimodal particle-size distribution of the crusher product, as coarse particles tend to roll down slopes of finer-grained material and accumulate at the bottom [24]. Besides, coning and quartering may become rather impractical beyond a certain bulk sample size. While in principle coning and quartering and/or knife edging are able to preserve representativity in a subsample, the method demands that procedures are applied rigorously, meticulously, and consistently.

The rotary-cone sample divider or splitter (aka. spinning riffler) is a popular laboratory alternative that greatly alleviates operator effort and vastly reduces inter-sample variation and thus analytical error. Dry bulk sample material is conveyed via a vibratory feeder off-centre on the surface of a rotating cone made of steel, aluminium or ceramic in a continuous stream and at a controlled rate. The spinning cone evenly divides falling particles over sample containers that are directly attached to the rotating dividing head. The apparatus closely resembles a merry-go-round in appearance. Slightly larger equipment versions instead rotate a chute over subsample containers fixed along a cone base, material not collected by the subsample containers (i.e., the largest portion) is recollected as reject in a separate bin. Larger equipment often uses a carrousel of sample bins rotating under a feeder funnel (see Fig. 3.2), and is capable of handling up to several tons of material and/or very coarse material, also outdoors in a quarry.

The actual sample division process is automated requiring no operator attention or experience, which again saves time and minimizes inter-sample variation

Fig. 3.2 Large-scale spinning riffler for splitting bulk sample materials. Reproduced from Fig. 2 in [81]

(provided all preceding steps have been correctly implemented). A broad range of cone divider models and sizes, in a variety of design variations, is available from many manufacturers to suit a laboratory's specific needs.

Homogenization of the sample is necessary e.g., to prevent or correct segregation (see next section on comminution artefacts), to ensure homogeneity when joining subsamples (of identical particle size), or when blending with an additive (e.g., powder flux for digestion - see separate section). Homogenization of a single sample can be done manually by pouring the material on a cone onto a clean tray or piece of kraft paper, flattening it, and then piling up on a cone again from the cake periphery using an appropriately sized trowel or shovel, and reiterate as necessary. However, this is labor intensive and subject to operator experience (and error), in the same manner as coning and quartering is (see above). Therefore, it is strongly recommended to use mixing equipment.

Larger sample volumes, especially from coarse particulate materials like gravel, are conveniently homogenized and mixed in a concrete mixer, which is available in a variety of sizes. Most of these mixers rely on the tumbler principle, assisted by fixed steel blades inside the drum. Models for designed for laboratory use commonly tend to have more precise speed and time controls compared to their onsite building and construction counterparts. Smaller sample volumes and/or of finer-grained materials can be effectively homogenized and mixed with a mortar mixer, in which the sample is typically agitated with a whisk or hook, like e.g., a dough mixer at a bakery or in a kitchen.

There are spiral mixers where the bowl rotates under the spinning agitator, and planetary mixers where the spinning agitator describes a planetary trajectory inside a fixed bowl.

Whether by blades fixed to the drum wall or by an agitator, the intense contact of the equipment parts with an abrasive sample like aggregate is likely to contaminate the material (even though some vendors offer polypropylene or polyethylene liners). For mixing concrete or mortar to be poured into a mould, contamination from equipment is hardly relevant. However, this is different for chemical assay, as already discussed above, especially when the size of subsamples after comminution and splitting is only a small fraction of the original bulk 'as received' sample.

For that purpose, mixing equipment with minimal contamination of sample material has been developed. The subsample material is put in a vial and gently rotated, inverted, and translated, to ensure thorough mixing and homogenization, with minimal agitation. The motion of the sample vessels is called 'turbula', after the geometric principle of the tumbling apparatus of Schatz [209]. The sample vessel is suspended between two rotating convex rockers laying in different planes (imagine a can of beans: one rocker is connected to the top in N-S direction, the other rocker is connected to the bottom in E-W direction). Simultaneous rotation of both rockers in opposite sense results in the sample vessel (bean can) describing a complicated tumbling (so-called 'oloid') movement, ensuring excellent mixing of contents. The only contact of the sample is with the container wall, whose material (glass, stainless steel, HDPE, PTFE) can be chosen to interfere minimally with the sample composition.

Turbula-system mixing and homogenizing equipment is available from a number of suppliers, in a range of capacities to suit specific demand. Especially the smaller versions are able to handle multiple samples at once, which is convenient for subsamples for analysis. Alternative systems certainly exist, but are not elaborated in further detail here.

As with comminution, thorough cleaning of all equipment parts in contact with sample material (e.g., cone, chute, bins, vials) is essential. Well-designed equipment facilitates cleaning and avoids deposition of sample material, as well as it limits parts wear.

3.2.1.3 Final Fineness of the Pulverized Product

To reinstate, proper attention was paid to ensure that the original sample as submitted was representative, and that its representativity is preserved during subsequent comminution, to prepare the material for analysis. The exact weight needed of a (sub-) sample for analysis is primarily determined by the analytical method and/or instrumentation of choice, generally on the order of a few hundreds of milligrams for analysis of 'main rock-forming elements' including ignition loss (LOI), total (organic) carbon/sulfur/nitrogen, as well as chloride and fluoride. Test portion powders are converted by flux fusion melting to form a glass bead for main element analysis by XRF, or dissolved in concentrated strong acid for analysis by ICP.

However, trace element analysis by XRF instead uses specimen produced by blending sample powder 'as is' with a specialized organic wax and pressing this into a pellet [29, 125]. Similarly, XRPD analyses the powder 'as is' loaded into a specimen holder, as any form of digestion is destructive to crystallinity. In both these methods, the analysed volume represents only a small fraction of the total test portion, as the particles 'in diffraction' (see e.g., [217], section "III. Conditions for success") or 'in fluorescence', respectively, are necessarily located near the prepared specimen surface (to avoid contamination of the analytical result by the sample holder).

That means that the actually analysed volume is only a small fraction of the overall subsample representing the test portion. Therefore, the statement by Injuk et al. ([125], pp. 412) cannot be overestimated: "The desired end result of specimen preparation is therefore to make sure that the relatively small volume of specimen analysed (…) is representative of the bulk".

Thus, the ultimate test portion of a few hundred milligrams is subject to the same mathematically determined particle-count statistics as all intermediate stages of preparation are, including subsamples of comminuted material, as well as the final powder. Below figure presents a diagram prepared cf. Gy's mathematical model on sampling uncertainty. The modal content of a sample constituent (i.e. 'species of interest' - SOI) is plotted along the ordinate axis on the left on a ^{10}log scale, and the nominal number of particles that the sample must contain to be representative for the particular SOI is plotted along the abscissa at the bottom.

First, the desired relative error level is selected in percent (in the below diagram at 2%-relative indicated by the red curve), e.g., as defined in a guideline or standard.

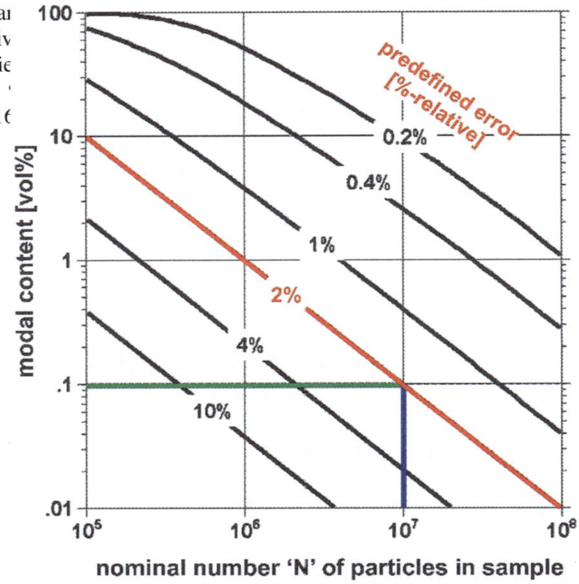

Fig. 3.3 Nominal number of pai cles needed in a subsample at a giv modal content in vol%, to achie a predefined accuracy level in ' relative. Modified after Loubser [1(

Second, the modal content in volume percent (vol%) of the species of interest SOI is located along the ordinate axis, and a horizontal line drawn (as indicated in green). The required number of particles in the (sub-) sample is now read on the abscissa (10^7 particles, i.e. ten million) right below the point of intersection with the curve for the predefined error, as indicated by the blue vertical line. While ten million particles indeed is a large number, it is three orders of magnitude (i.e., 10^3, or 1000×) less than typical particle numbers in adequately prepared powder diffraction samples (see [217], 186ff) (Fig. 3.3).

The abundance (0.10 vol%) and error level (2%-relative) shown in the diagram above would give an accuracy within 0.10 ± 0.002 wt%, typical for e.g., apatite as the main phosphorus mineral in bulk igneous rock. Such low numbers do not normally apply to potentially ASR-deleterious constituents in concrete aggregate, where (alkali-reactive) species of interest are commonly in the range of 2.0 vol% (e.g., chert in Dutch gravel) to 20 vol% (e.g., mylonite in Norwegian aggregate). However, rapidly mounting experience with deleterious expansion caused by pyrrhotite (=iron sulfide; FeS_{1-x}) present in concrete aggregate (reused tailings from metals mining, or tunnel blasting) does in fact comprise such low concentrations (see e.g., [201]). Regardless of actual modal content in vol% and/or desired error level, the principle of representativity across all stages of sample preparation (comminution, splitting, homogenizing) remains intact and in full effect.

As the weight of the test portion (and/or subordinate 'analysed volume', as applicable) is known, the final particle size (fineness) can be calculated from the weight of the test portion (or the 'analysed volume' subset, respectively), divided by the

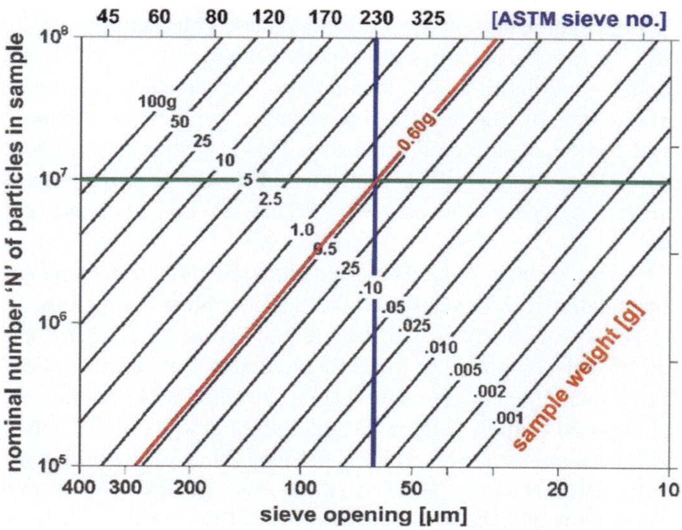

Fig. 3.4 Particle fineness needed to obtain a nominal number of particles in a subsample of given weight. Modified after Loubser [162]

number of particles needed. With that number, the test portion *s.l.* will remain statistically representative of the 'as received' sample. Using the second diagram shown above in Fig. 3.4, the previously determined nominal number of particles (i.e., 10^7) needed is first located along the ordinate (vertical axis), and a horizontal line drawn (green in the diagram). At the point of intersection with the curve for the relevant test portion weight (as defined by analytical method and instrumentation, in the example here set to 0.60 g highlighted in red), a vertical line is drawn to read the particle size (fineness) required in micrometer (bottom abscissa, ~65 μm), or in ASTM sieve numbers or 'American mesh' (top abscissa, 230 mesh).

Thus, in summary, a sample material must be pulverized to pass a 230 mesh sieve to achieve a final fineness <65 μm, which ensures that the test portion (or 'analysed volume', respectively) nominally comprises ten million (=10^7) particles. This number warrants that the sample material contained is statistically representative for the species of interest SOI (with a modal content of 0.1 vol%).

3.2.1.4 Liberation of Aggregate from Hardened Concrete

Occasionally, it is desirable to extract aggregate material from hardened concrete for assessment. This is not an easy task, and several methods have been tried, with variable degrees of success. In principle, material can be liberated mechanically by comminution, or chemically by selective dissolution. Usually, aggregate is extracted from concrete samples of limited size, e.g., drilled cores of say Ø100 mm × 400 mm, weighing ~7.5 kg, containing about 1.25 kg binder, 2.50 kg fine aggregate, and

3.75 kg of coarse aggregate (simple 1:2:3 concrete). Thus, samples of this size are *not* representative for coarse aggregate, except if monomict and fine grained.

Conventional comminution equipment (e.g., jaw breaker, hammer) diligently reduces particle size of all concrete constituents, regardless hardness, tenacity, friability, particle size, or (mineral) grain size. This is desirable for bulk whole-rock chemical analysis (ignoring artefacts as discussed above) yet renders the equipment unsuitable if the aggregate is to be preserved 'as is', i.e., liberated only, without comminution.

Non-mechanical 'electric-pulse disaggregation' (EPD) is a relatively new method, which has been extensively investigated for the liberation of precious and delicate ore minerals, gems or fossils from hard rock (e.g., [1, 148, 208, 256]). For a historical overview and time-development of the EPD method and equipment, see [16].

For EPD treatment, the concrete core is first saturated with water, then immersed in water and exposed to high voltage pulses. Pulses below ~80 kV mostly decompose water, whereas pulses above ~100 kV produce shock waves within the rock that disintegrate the material along 'zones of weakness', mostly mineral grain boundaries. Even though the physico-chemical details of EPD are still not fully understood (including plasma-formation), with today's equipment it is possible to control the pulse shape (rise/decay times) and duration, repetition rate (~ 1–5 Hz), and total amount of energy delivered (kJ), to optimize the effect for the material at hand. The efficacy of EPD for the liberation of aggregate from (damaged) concrete has been demonstrated [16, 268] showing stunning selectivity between paste and aggregate. Therefore, the EPD method is superior to the alternatives outlined below. Laboratory EPD equipment designed for batch processing is commercially available from several vendors.

Rather than using mechanical force, aggregate might be liberated using chemicals that only attack the paste binder. Hydrous solutions of strong acids (e.g., perchloric $HClO_4$, hydrochloric HCl, sulfuric H_2SO_4, nitric HNO_3) or mixtures (e.g., *aqua regia* $3HCl{:}1HNO_3$, a rather common mixture in geochemical laboratories) are used to digest rocks and minerals, whereas digestion of silicate minerals requires HF (or a cocktail like $5HClO_4{:}3HNO_3{:}2HF$). Concentrated solutions of these (and yet other) acids and/or cocktails are popular in routine procedures for bulk whole-rock analysis because they readily dissolve common rock-forming minerals with little residue (see e.g., [60, 192]). In other words, strong acids (mixtures) are pronouncedly non-selective, and therefore immediately unsuitable for the selective dissolution of paste in order to liberate the aggregate [91–94], though exceptions have been reported [24, 25].

Solutions of weak organic acids like acetic acid (H_3C-COOH, 'ice-vinegar'), oxalic acid (HOOC-COOH, 'di-acetic acid'), formic acid (H-COOH, 'ant acid'), or weak mineral acids like phosphoric acid (H_3PO_4) and boric acid (H_3BO_3, mineral name: sassolite) are widely used to remove unsightly mineral encrustations from museum or collectors specimens (see e.g., [67, 135, 136]).

One alternative method to liberate aggregate from hardened concrete first freezes water-saturated concrete (cores) in liquid nitrogen at −196 °C, then shock-heats the deep-frozen cores in a microwave oven (see e.g., [109]). The freeze–thaw treatment is reiterated until the concrete crumbles, which is further promoted by 'tapping lightly with a hammer'. The number of freeze–thaw cycles needed to fully disintegrate the composite material depends on the general constitution of the actual concrete, as a net sum of nominal strength, binder-to-aggregate bonding, presence or absence of pre-existing cracks, overall porosity and permeability as well as distribution in the sample volume, etc.

The thermal cycling produces extensive cracking from expansion of pore water upon freezing to ice, thermal stress upon heating from anisotropic expansion, as well as from internal pressure from steam development.

However, aggregate particles commonly have adhering paste and are thus only partially or poorly liberated. Paste may be removed by rinsing with nitric acid (cf. [109]). However, strong acids will also readily dissolve any carbonate minerals that the aggregate might contain, thus attributing a preparation artifact of unknown and variable extent (because of inter-sample variation). Therefore, this freeze–thaw method is unreliable and thus not recommended.

3.2.1.5 Comminution Artefacts

Obviously, comminution is a destructive process that obliterates the original microstructure, texture, and fabric of the material even down to crystal-structural scale (unfit for XRPD). If a powder is prepared for chemical assay, then the maxim 'the finer, the better' applies. The tenet of 'collective properties-preservation' applies in principle to the sample bulk at all stages of analysis: the 'as received' bulk, the crushed product, the subsample split, as well as the final powder for specimen preparation. The exact constitution and quality of any individual powder particle is irrelevant - it is the collective of particles that together represents the material and its (geo-) chemical properties.

In contrast to (geo-) chemical assay, however, aggregate petrography actually relies on microstructure, texture, fabric, and mineral content for identification of rock type, in addition to particle counting for modal analysis. An 'as received' bulk sample may have to be comminuted prior to analysis, to increase the number of particles assessed to fulfil statistical requirements (see above; also see [27]).

Consider a representative sample (50 kg of ASTM D75 [2]) of nominal 25 mm virgin aggregate particles of crushed granite, composed of 5 mm mineral grains of reddish K-feldspar, white oligoclase plagioclase, colourless quartz in equal volume proportions (32 vol% each) in a random granoblastic fabric, with minor (4 vol%) black biotite. The bulk material is strictly monomict, as immediately observed from the 'as received' sample. Comminuting this material by crushing in a jaw breaker to 10 mm particles will not change much. However, as soon as the particle size of the crushed product approaches mineral grain size (5 mm) or finer, the originally monomict material suddenly transforms into a polymict material, with particles consisting exclusively of K-feldspar, oligoclase, or quartz, sometimes with biotite. Only a minority of particles will still consist of 'granite' with all four minerals. How many of such ‘clue particles’ will end up in thin-sections for petrographic assessment cf. RILEM AAR-1.1? Without knowledge of the parent sample, the comminuted material will be reported as polymict, which in fact represents a preparation (=comminution) artefact. Nota bene: if the same granite were also used to manufacture sand by crushing it to <2 mm, then reporting the material as polymict does not comprise an artefact, because the sample ‘as received’ *is* already polymict.

In addition, liberated monomineralic grains are now also prone to gravity segregation. While the densities of K-feldspar (orthoclase: ~2.60 $kg{\cdot}L^{-1}$), oligoclase (~2.62 $kg{\cdot}L^{-1}$), and quartz (~2.66 $kg{\cdot}L^{-1}$) are quite near, other common minerals may be substantially heavier (e.g., biotite: ~3.00 $kg{\cdot}L^{-1}$, hornblende: ~3.35 $kg{\cdot}L^{-1}$, grossular garnet: ~3.59 $kg{\cdot}L^{-1}$, zircon: ~4.62 $kg{\cdot}L^{-1}$), whereas ore minerals (e.g., magnetite: 5.18 $kg{\cdot}L^{-1}$) and in particular native metals (e.g., copper: ~8.95 $kg{\cdot}L^{-1}$, gold: ~19.3 $kg{\cdot}L^{-1}$) are heavier still. Minerals heavier than the bulk average tend to concentrate at the bottom during comminution, splitting, and preparation in general. Obviously, lighter species rather concentrate towards the top, both processes leading to gravity segregation. While most so-called ‘heavy minerals’ >5.0 $kg{\cdot}L^{-1}$ normally have low abundant concrete aggregate in general (except, e.g., magnetite), any particulate material containing particles with significantly different densities is potentially prone to gravity segregation. Segregation can be minimized by following rigorous routine in material handling and processing, by using suitable equipment, and by homogenization (see preceding section) of the (sub-) sample before each next processing step.

As another example, consider a granitic gneiss with light coloured bands of granitic mineral composition (see above), alternating with dark bands containing biotite and amphibole. In the field, the banding is seen to vary in thickness from a few centimetres to several decimetres. A 50 kg sample of nominal 25 mm virgin aggregate material from such a gneiss will be designated as polymict, consisting of granitic and amphibolitic particles. Only a few particles will consist of light–dark banded gneiss, but most will be either light (granitic) or dark (amphibolitic), not both. Here, field information is essential to identify the material as a derived from a single lithology (see §3.1 in RILEM AAR-1.1, pp. 38). Additional comminution may reduce the proportion of ‘gneiss particles’ further.

Finally, consider igneous rock types with a porphyritic texture of e.g., coarse-grained K-feldspar phenocrysts in a much more fine-grained matrix of K-feldspar (different population), quartz, plagioclase, and additional minor and trace minerals. The rock has an all-natural bimodal grain-size distribution. Comminution of such a rock to particles smaller than the interspacing distance of individual phenocrysts will also produce particles consisting of the fine-grained matrix only, without the porphyritic texture of their bulk rock parent. This is again a comminution artefact that is challenging to interpret correctly, especially if field information is lacking. The same artefact will also affect metamorphic rocks with coarse-grained porphyroblasts in a fine-grained matrix (e.g., blasto-mylonite matrix fragments, augen-gneiss). Even some sedimentary rocks (e.g., matrix-supported conglomerate) will experience the same artefact when particle size upon comminution becomes less than the interspace distance between pebbles floating in the matrix.

In addition to the above 'liberation' comminution artefacts, either a bimodal particle size distribution, or the presence of non-cubic particles (i.e., one orthogonal axis being significantly longer than the other two-acicular particles, or shorter-flaky particles) are also undesirable, as both qualities promote segregation by size (smaller particles falling down between larger - the 'potatoes in a bag' effect), or by shape (elongate and flaky particles roll differently from cubic particles). Thus, they affect further processing and impart reliable splitting into subsamples. Both bimodality and/or non-cubicity of the crushed product must be corrected immediately after they have been recognized, and both demand homogenization of the corrected product.

The above serves to illustrate that correct interpretation of analytical data (geochemical, petrographical) demands combining observations from a range of size scales. Generally, geologists are well aware of this issue, as they are familiar with studying the largest objects on Earth (i.e., 'Earth') while only being able to analyse samples of limited size.

3.2.2 Digestion of Pulverized Sample Materials for Analysis

Sample acquisition and comminution to prepare the material for analysis have been elaborated above. The required weight of pulverized sample material and its ultimate particle size/fineness are important prerequisites for reliable analysis, with reproducible results. This section describes the preparation of pulverized sample material into specimen for analysis. In addition to methods for bulk whole-rock geochemical analysis, selected specialized methods for specific purposes are described as well.

The main constituent elements in geological materials (e.g., rocks, aggregate, cement, SCMs) with an analytical sum total equal to (somewhat arbitrary) 95 wt% or more are commonly identified as the 'main' or 'rock-forming' elements, and are typically listed as oxides. These also comprise volatile species like H_2O, CO_2 (TC; 'inorganic carbon' from carbonate CO_3^{2-} orgraphite), 'total organic carbon' (TOC), total sulfur (TS; SO_3), and more rarely also 'total N' (TN; i.e., nitrate NO_3^-, ammonium NH_4^+), as well as chloride (Cl^-) that are normally analysed on separate sample aliquots, in separate procedures.

In the early days of geochemistry, each constituent element in the sample material was separated by chemical manipulation and weighed individually on an analytical balance, making comprehensive bulk whole-rock analysis extremely labour intensive and time-consuming. This is commonly referred to as the 'classical method of analysis'. With the advent of photometry, spectrometry, and the accelerating development of automated equipment and instrumentation, this has gradually evolved into 'rapid analysis'. Modern equipment for sample and specimen preparation, as well as for feeding ready-prepared specimen into analytical instruments have improved reproducibility by removing inter-operator variation, and have vastly reduced time and effort, and through that, lower cost per analysis (e.g., [115], also see [192]).

This STAR-report describes the most widely used methods for 'rapid analysis' of bulk whole-rock composition (i.e., XRF or ICP, plus LOI and total S/total C) and typical procedures for sample and specimen preparation. In addition, two specific methods for assay of rock constituents are described.

It should be well-noted that some minerals have refractory properties, meaning they exhibit a high corrosion-resistance against molten alkali salts ('fluxes') and/or concentrated strong acids. Well-known examples include quartz (α-SiO_2, an oxide mineral) and silicate minerals (i.e., containing $(SiO_4)^{4-}$ groups) in absence of HF (though many silicates tend to dissolve incongruently in aqua regia), alumino-silicate minerals andalusite, kyanite, sillimanite, mullite and crystal-structurally related staurolite, spinel-group minerals especially those rich in Ni, Cr, and/or Ti (e.g., 'magnetite' with common impurities), as well as zircon, corundum, tourmaline, beryl, cassiterite, and a number of REE, U and/or Th bearing minerals (see e.g., [29, 50, 60, 192, 230]). This brief listing is by no means complete; many more refractory minerals have been identified. Therefore, it should be clear that, once more, thin-section petrography to determine mineral (modal) content is essential for the selection of the best-suited methods of specimen preparation and analysis, to warrant reliable results.

Local laboratory protocols may differ in details from the rather generic descriptions given here, to accommodate for specific laboratory hardware including instrumentation, in-house experience on specific sample materials or to answer to specific analytical demands.

Nota bene: by default, all chemicals referred to below shall be of highest purity grade, identified as e.g., Pro Analysi, Suprapur, Trace Metal Grade, or other. Water shall have a quality of 18 MΩ-cm or better. Where use of lower-grade chemicals of lesser purity is permissible, this will be specified.

3.2.2.1 Elemental Analysis by X-ray Fluorescence (XRF)

Assessment of the chemical compositions of geological materials is commonly done by X ray fluorescence (XRF) to determine contents of main elements (as oxides), together with loss on ignition (LOI) to determine contents of volatile species. Modern XRF instruments are capable of analysing twelve or more elements in one single assay, usually including Na, K, Mg, Ca, Mn, Fe, Al, Si, Ti, and P, with typical lower-limits of detection (LLD) values on the order of ~0.01 wt% for each element (Na usually 0.10 wt%). Lower detection limits can be achieved by press-pellet preparation (see below). Analysis of Ca, Mg, and Al in carbonate rocks as prescribed by Annex A in RILEM AAR-0 [198], pp. 19) can be done by XRF as well. Today's instruments are nearly always equipped with a robot to load specimen into the instrument for analysis, enabling unattended operation outside normal office hours.

The method relies on fluorescence of elements present in flux-fusion bead or a press-pellet, upon exposure to X-rays generated by a known source, commonly with a Rh-target. The fluorescent spectrum is recorded by a number of wavelength-dispersive spectrometers, and digitally processed to derive element contents. For bulk whole-rock analysis, these contents are normally converted to oxide contents in wt%, assuming stoichiometry of each oxide. Iron may be presented as FeO or Fe_2O_3, with the connotation '-total' as XRF is unable to assess oxidation status.

Concise introductions to XRF analysis are given in e.g., Sibilia [214] and Leng [152], in-depth detail is provided in e.g., Van Grieken and Markowicz [249] and Beckhoff et al. [6]. Whole-rock analysis by XRF is a routine procedure in geology/ geochemistry and environmental characterization, and is offered at competitive cost from accredited suppliers.

Data shall be presented in a table together with net LLD (LoD) values per element oxide species, and must include analytical sum totals. The net cumulative sum of rock-forming oxides in a well-performed bulk whole-rock analysis by XRF, supplemented with ignition loss LOI and total C/total S (as applicable) should be within 100.0 ± 1.50 wt%. Results with sum totals outside of ± 1.50 wt% should be discarded and the analysis redone. Element concentrations lower than ~5–10 × the respective LLD are generally considered of indicative value only, unsuitable for modelling.

3.2.2.2 Elemental Analysis by Inductively-Coupled Plasma (ICP) Spectrometry

A common alternative to whole-rock analysis by XRF is analysis by inductively-coupled plasma—atomic emission spectrometry (ICP-AES), sometimes also referred to as ICP-OES for optical-emission spectroscopy. The method relies on atomic/ optical emission of elements dissolved in a hydrous solution, nebulized into a high-temperature (argon—Ar) plasma at up to ~10,000 K generated and maintained by an induction coil. The emitted spectrum is recorded by axial and/or radial spectrometers, and digitally processed to produce element contents in wt%. Modern ICP-AES/OES instruments are capable of analysing 40 or more element species in a single assay, and lower limits of detection (LLD or LoD) in routine analysis are typically on the order of parts per million (1 ppm = 1 $\mu g \cdot g^{-1}$).

Analysis by ICP-AES/OES is elaborated in e.g., Varma [252] and Potts [192]. Whole-rock analysis by ICP-AES/OES is a routine procedure in geology/ geochemistry and environmental characterization, and is available from accredited suppliers at competitive prices.

More and more laboratories are replacing and/or upgrading their current atomic/ optical emission spectrometer (ICP-AES/OES) instrument with a mass-spectrometer (ICP-MS). Generally, this newer instrument type has lower limits of detection, whereas sample preparation procedures remain unchanged.

3.2.2.3 Combustion Analysis of Total Sulfur (Total-S, TS) and Total Carbon (Total-C, TC)

A common supplementary method is combustion analysis for elements that may be volatilized during flux-fusion for XRF or acid-digestion for ICP, most notably carbon C and sulfur S. The combustion method relies on volatilization of all carbon and all sulfur of undiluted dry raw sample powder in an oxidizing atmosphere at 1400 ± 50 °C. Carbon- and/or sulfur-containing minerals are decomposed releasing C and S, which are measured as oxides CO_2 and SO_2 in the effluent gas by infrared light absorption. Modern instrumentation allows simultaneous analysis of total-C (or TC) and total-S (or TS) on a single aliquot.

Combustion analysis is a routine supplement for whole-rock analysis by either XRF or ICP, and is quite commonly referred to as 'Leco analysis' after a main instrument manufacturer. Combustion analysis ('Leco') is commercially available from accredited suppliers at competitive prices.

3.2.2.4 Gravimetric Analysis of Loss-On-Ignition (LOI)

A further common supplementary method is gravimetric analysis of ignition loss. Heating of an aliquot of undiluted dry raw sample powder at ~1050 ± 50 °C releases volatile species present in the material. The weight difference before and after heating is recorded as 'loss on ignition' (LOI), and normally includes all hydrous species as well as organic and inorganic carbon, whereas sulfates typically are not affected [234]. It should be noted that the recorded weight loss potentially can be partially compensated by weight gain through oxidation of metals or metallic compounds, e.g., iron or ferrous iron Fe^{2+} to ferric iron Fe^{3+}. Ignition loss is essentially a default supplement completing whole-rock analysis by XRF and/or ICP, and is commercially available from accredited suppliers at competitive prices.

3.2.2.5 Sample Preparation for Analysis by XRF Flux Fusion

To annihilate any remaining heterogeneity, mineralogical, or particle size effects (**NB**: unrelated to representativity, already considered in earlier preparation steps), the test portion is diluted and fused at high temperature with a fluxing agent to produce a homogenous glass bead.

In a typical procedure, an amount of powdered sample material is pre-dried at 105 °C to remove absorbed water, precisely weighed (e.g., 0.600 ± 0.005 g) with 3-7x excess by weight (e.g., 4.200 ± 0.005 g) of flux powder, and thoroughly mixed in a turbula-type mixing apparatus. The fluxing agent typically consists of lithium metaborate ($LiBO_2$) or lithium tetraborate ($Li_2B_4O_7$), or a mixture thereof, because of its low X-ray absorption and excellent fluxing properties, as well as that both lithium (Li) and boron (B) do not interfere with most common whole-rock compositions. Certain elements are known to form volatile compounds at high temperatures and will thus escape during fusion, notably in samples with high fluoride and/or chloride contents (also see section on acid digestion for ICP analysis). However, a broad choice of alternative fluxes is available for special applications (Table 2.5 in Potts [192], p. 53, also see [29, 60, 230]).

The mix is poured into a crucible and digested at ~1050 ± 50 °C, transforming into a clear transparent liquid without bubbles. The melt is poured into a polished die and allowed to cool, forming a ~5 g glass bead with a flat and high-gloss surface (to minimize scatter) for analysis. High-purity platinum alloyed with 5wt% Au (95Pt05Au) is a preferred crucible material, due to its suitably high melting point at 1772 °C, non-oxidizing and non-wetting behaviour so that the melt is poured out completely and the bead effortlessly released from the die, respectively. Use of alternative fluxes may require use of alternative crucible materials [29, 192].

The flux-melting procedure is referred to as the 'Claisse method' after its inventor [45], also see [29, 125]. Fully automated equipment is available taking care of melting (by gas flame or induction), mixing and homogenizing of the melt, pouring into the die, and cooling at a controlled rate, alleviating operator effort while enhancing reproducibility. The flux-fusion method's suitability for a broad range of geological and related materials is extensively documented (e.g., [50, 161, 182], and references therein), though some minerals are known to stubbornly resist [60, 230].

For analysis of trace elements in low concentrations, sample powder is mixed directly with a weighted small amount of an organic binding agent, then mechanically pressed into a pellet. Common binders include potato starch, ethyl cellulose, alcohol, urea, lucite, and water, which all consist of light elements (H, C, N, O) with low X-ray absorption [29]. As the sample material in the press-pellet specimen is much less diluted compared to flux-fusion, generally lower detection limits can be achieved, making the press-pellet method suitable for trace element determinations. On the other hand, dilution might be required to avoid detector overexposure when measuring abundant element species, so preparation of press-pellets is not always an alternative for flux-fusion.

For very large numbers of samples, comminution and digestion are very time-consuming as well as labour-intensive, and consequently become cost-prohibitive. To circumnavigate these aspects, Van Gaans et al. [247] developed an alternative method called "integral rock analysis (IRA)", which was field-tested during a large sampling campaign with positive result. Bulk rock samples collected in the field are core drilled in the laboratory with a diameter to precisely fit an XRF instrument specimen holder. Next, the cores are entirely sliced up to accurate thickness with a diamond blade, and all slices polished using automated polishing equipment, saving tremendously on tedious labour and precious time for crushing, pulverizing, splitting/homogenizing, as well as equipment cleaning between steps.

Finished slices are mounted in a specimen holder and loaded directly into the XRF instrument for analysis. The cumulative result of the individual slices in a single drill core can be used to approximate bulk whole-rock analysis by ordinary procedures. For small samples with relatively short drill cores, slices can be polished and analysed both sides. In addition, any individual specimen giving an unexpected result can be inspected in an ore microscope in reflected light or with SEM–EDS, to directly identify any unexpected mineral phases present. Latter option is not possible with traditional preparation methods as elaborated above, as pulverization and digestion (flux-fusion) obviously are destructive.

For further detail on digestion of sample powders by high temperature fusion in alkali salts, press-pelletizing, or for preparation of non-solid sample materials, more information is given in [29, 50, 60, 125, 161, 183, 192, 247] and references therein.

3.2.2.6 Sample Preparation for Analysis by ICP: Digestion in Acid

For analysis by ICP, the powdered sample material needs to be brought into aqueous solution to be nebulized into the plasma for analysis by AES/OES or MS. Non-silicate minerals (e.g., carbonates, phosphates, oxides) are readily decomposed in concentrated strong acid, e.g., hydrochloric (HCl), perchloric ($HClO_4$), nitric (HNO_3), or a mixture thereof, e.g., aqua regia ($3HNO_3$:1HCl) which also dissolves gold (hence its common name, Latin for 'King's Water'). In contrast, silicate minerals only fully decompose in hydrofluoric acid (HF), forming the hexafluoro-silicate ion (SiF_6^{2-}) that is stable under these conditions (see e.g., [41, 42, 192], and references therein).

> **Nota bene**: hydrofluoric acid classifies as an EXTREMELY HAZARDOUS chemical, demanding use of full protective attire (full-face visor, long arm gloves, full-length apron, boots) as well as a polymer-lined fume hood (glass will be etched).
>
> MAKE SURE TO COMPLY WITH ALL SAFETY MEASURES BEFORE USE.

Technically speaking, hydrofluoric acid is not a 'strong acid', meaning that the HF molecule is not fully dissociated in concentrated aqueous solution (contrary to e.g., the strong acid HCl), but rather more like the also weak acetic acid (H_3C-COOH). However, fluoride salts of many main rock-forming elements are only sparingly soluble in a hydro-fluoric acid solution, especially calcium (precipitates as fluorite, CaF_2) and magnesium (precipitates as sellaite, MgF_2), among others [18, 146]. Therefore, hydrofluoric acid is hardly ever used by itself, but usually mixed with other acids (e.g., $5HClO_4$:$3HNO_3$:2HF or $3HNO_3$:1HCL:1HF), to prevent unwanted precipitation of insoluble species. The low solubilities of many common sulfates and phosphates prevents practical use of sulfuric (H_2SO_4) and (ortho-) phosphoric acid (H_3PO_4), besides some specialized applications (see below).

After dissolution in excess concentrated strong acid, a number of elements tend to form volatile compounds with fluoride and/or chloride, and are therefore prone to evaporation from the solution, as specified in Table 2.2 of Potts ([192], pp. 48). Silicon in solution can be stabilized effectively by the addition of boric acid (H_3BO_3) (see e.g., [192, 252]), which also somewhat ameliorates the hazard potential of a fluoride solution, though utter care remains obligatory.

Alternatively, the silica (and fluoride from solution) can be lost deliberately, which has several advantages. Solutions without silica suffer much less from aging (from silica polymerization), and also suffer less from potential interference with other species in the analyte. The overall salt content of the analyte is substantially reduced, resulting in a lower analytical background. The purging of fluoride also annihilates the corrosive effect of HF, so that further manipulation can be done with regular glass labware. That includes the glass tubing in the interior of the ICP instrument, which normally needs to be changed to corrosion resistant aluminium oxide tubing to handle HF-solutions. The multiple advantages of getting rid of silica and fluoride have made this alternative arguably the most common digestion practice.

The choice to lose (or retain) silica plus fluoride (as well as other element species) must be made prior to digestion, as procedures and chemicals for either option are entirely different. Two methods exist, both losing silica (and other elements, as elaborated above) by volatilization.

In a typical procedure, 0.500 ± 0.005 g of pulverized sample material is weighed into an open Pt beaker, to which is added 5 mL of concentrated perchloric acid (60 vol%), and then 15 mL of 40 vol% hydrofluoric acid, in that particular order to ameliorate the otherwise vigorous reaction with silicate. The beaker is heated to 265 ± 15 °C in a closed fume hood until incipient dryness, embedded in a sand-bath for optimal heat transfer. Any volatile species present will escape during this stage. Complete dryness is to be avoided, to prevent precipitation of insoluble salts (see e.g., [192]). This step may be reiterated as necessary to obtain satisfactory digestion (see [192], and references therein). Next, the residue is taken up in ~5 mL of 1N hydrochloric or nitric acid, and finally made op to 100 mL (at ~pH 2) using double-distilled water.

Alternatively, 1.000 ± 0.001 g of pulverized sample material is weighed into a PTFE (poly tetrafluoro-ethylene) or PFA (perfluoro-alkoxy) beaker with a self-sealing lid of the same material, and 20 mL of 7N HNO_3 nitric acid added. The beaker is placed in a pressure-resistant stainless-steel or Monel-alloy autoclave jacket, sealed and heated in an electrical furnace at 130 ± 20 °C (e.g., [147]). Instead of conventional heating, digestion can be assisted by microwave irradiation, requiring a microwave-transparent (polymer or ceramic) jacket rather than metal (see e.g., [141, 231]). Internal pressure in the beaker may rise to 7–10 MPa (70–100 bar). After completed heating (1–48 h), the autoclave vessel is removed from the furnace and allowed to fully cool down to ambient temperature before opening. 'Silica' *sensu lato* can be lost by subsequent heating of the opened vessel in a sand-bath as described above. The nearly dry residue is again redissolved in ~5 mL of 1N hydrochloric or nitric acid, then made up with double-distilled water to 100 mL. The very severe conditions during heating greatly enhance the solubility of minerals in general as well as accelerate their digestion. This procedure is colloquially referred to as (microwave-assisted) 'bomb digestion', as applicable. An extensive method review is given in Lamble and Hill [144].

3.2.3 Specific Methods of Digestion

3.2.3.1 Phosphoric Acid Method to Determine Free Silica Content in Rock Material

This method liberates crystalline silica minerals quartz, cristobalite, and tridymite from the rock by selective digestion of silicates, carbonates and oxides in the assemblage. The method was originally developed to determine the free silica content in airborne dust as a silicosis hazard, and has been adopted by the Japanese cement industry [128, 170, 171].

For liberation, a representative (sub-) sample is pulverized to pass a 75 μm sieve using clean laboratory equipment, and 0.30 g of dry pulverized sample is weighed into a 250 mL Cochran beaker (~neckless Erlenmeyer). Carefully add 25 mL of 15M concentrated hydrophosphoric acid (H_3PO_4) solution and cover the beaker opening with a glass Hirsch funnel to limit evaporation. Quickly bring to a boil using an electrical heater and simmer very gently for 12 min, swirling the beaker gently but regularly to avoid overheating. After boiling, put the beaker on a cold surface and remove glass funnel, letting adhering liquid run down along the flask wall. Allow to cool to ambient temperature, then wash beaker wall with 125 mL of 60–70 °C hot water and swirl vigorously to dissolve the viscous phosphoric acid completely. Do not shake to avoid uptake of air.

Next, in a fume hood wearing safety goggles and appropriate protective clothing (i.e., arm-long gloves), add 10 mL of 1.55M fluoroboric acid (HBF_4) solution and swirl vigorously. Add 10 mL of paper pulp suspension and wash the inside of beaker with 25 mL of distilled water, then stand for 1 h, and filtered over ash-free cellulose paper. Next, the residue is washed with excess hot 3M hydrochloric acid (HCl) solution, then with hot distilled water to remove all traces of acid. The cellulose filter paper is then transferred to a tared platinum (Pt) crucible, and gently heated to dry and char the cellulose filter paper, then ignited to maximum 800 °C to remove any remains from the cellulose filter paper. The crucible is allowed to cool in a desiccator and weighed.

The residue weight represents the sum content in quartz plus cristobalite plus tridymite. Their relative proportions can be determined by X-ray powder diffraction (XRPD). For sedimentary rocks and in particular argillaceous carbonate rocks, the presence of crypto- to microcrystalline quartz can be confirmed using scanning electron microscopy (SEM) on polished thin sections.

Additional details on the digestion procedure for the liberation of crystalline silica are given in Talvitie [233]. An alternative method for the determination of amorphous silica is given in Kreshkov et al. [142].

Nota bene: fluoroboric acid HBF_4 dissociates and releases hydrofluoric acid HF upon heating. Take the same HSE precautions as mentioned above for hydrofluoric acid.

3.2.3.2 Macroscopic Detection of 'Invisible Sulfide' in Rock Materials

This method assists in the identification of very small quantities of fine-grained sulfide minerals like e.g., pyrite, marcasite, pyrrhotite, that are difficult to identify in dark coloured lithologies like black shale and certain greywackes, both macroscopically and in petrographic thin section.

Prepare reagent solution by dissolving 3 g sodium azide (NaN_3), 3 g potassium chloride (KCl), and 1.25 g iodine (I) in 100 mL double distilled water. The solution has a shelf life of several months if stored in a brown bottle and not exposed to direct sunlight. In contact with any sulfide which serves as a catalyst, the dissolved azide instantly reacts with iodine to form dissolved iodide and nitrogen gas, which latter effervesces and forms bubbles. The reagent can be applied with a glass bar to unprepared rock specimen, powdered material, or even petrographic thin sections without a cover slip.

The presence of any effervescence should be verified with a 10× hand loupe, or under a stereomicroscope if available. An ordinary cover slip can be placed temporarily over a thin section wetted with reagent solution to make tiny bubbles merge and become visible. Thus, any sulfides present in amounts as low as 0.3 μg can be detected. The method can also be used to identify 'opaques' in (non-covered) thin sections as oxides or sulfides. Additional details are given in Feigl [72].

3.2.4 *Preparation of Thin and Polished Sections*

Professor William Nicol of Edinburgh/SC was the first to study thin cuttings of fossil woods, minerals, as well as fluid inclusions under an optical microscope, equipped with the 'Nicol prisms' he had invented to obtain polarized light [179]. Nicol's pioneering work was continued by Alexander Bryson who had inherited Nicol's instruments and preparations. However, it was Sir Henry Clifton Sorby of Sheffield/UK who, seeing fluid inclusions during a visit to Bryson in Edinburgh, realized the potential of microscopical petrography on thin specimen of rock. Back in Sheffield, he produced the first thin section in 1849 by carefully lapping down a chip of Ledbury sandstone with powdered emery (=rock-forming corundum), mounted on an object carrier glass with Canada balsam. Today, Nicol and especially Sorby are recognized as the founders of optical petrography using thin sections. For additional detail on the early days of thin section petrography, the reader is referred to Chapter XV in Geikie [86].

In current petrographic practice, thin and/or polished sections may be used for diverse purposes:

1. Assessment of mineral content, microstructure, texture/fabric, and the like,
2. Assessment of porosity, voids, domains, microcracks, intragranular cracks, and the like,
3. Quantitative determination of constituents,
4. Primary assay in preparation for assessment by subsequent supplementary techniques,
5. Preservation of the present status of the sample material for long-term storage.

While one single thin/polished section in many cases may be able to cater more than one purpose, some purposes may be mutually exclusive (e.g., pt 4 and 5). Below subsections provide some background for selected aspects related to preparation of thin/polished sections. Generic descriptions of equipment, machinery, tools, and consumables are given in Weatherhead [261] and Humphries [123]. Hydraulic (e.g., clinker, cement) and/or water-soluble materials (e.g., rock salt) require special anhydrous preparation procedures for thin sections, see e.g., Spiers et al. [220], Engelhardt [70], Wang et al. [259].

Step-by-step instructions on how to proceed with the production of thin sections are given in manufacturer user manuals of the respective equipment and machinery.

3.2.4.1 Section Dimensions and Thickness

The 'standard format' for petrographic thin sections is the 'Gießener Format' at 28 mm × 48 mm named after the German University (town) where it was first put into practice [197]. The main manufacturers of optical or electron microscope instruments (including electron-probe micro-analysers - EPMA) worldwide all provide sample/specimen holders of 'standard petrographic size' (though exact dimensions may vary within ±1 mm), and wooden or cardboard storage boxes or transport carriers accommodating 'standard petrographic slides' are readily available. We will therefore assume this format as the 'standard size' for petrographic thin or polished sections. Carrier glasses are available from many suppliers in thickness ranging from 1.0–2.5 mm, in quartz or soda-lime glass. Thicker slides are more robust than thinner, an advantage if sections are taken along when traveling. Single-sided frosting enhances adhesion of mounting media; frosting by chemical etching is optically less conspicuous.

Some laboratories produce sections on a carrier glass of 50 mm × 75 mm (or even larger) to contain more particles of a given size, which is great for representativity (see e.g., [27, 119]). Other than the practical limitation of (standard) specimen-to-holder fit and available space in the microscope instrument, there is no technical reason against use of larger thin and/or polished sections.

The net accessible specimen area on a standard-size petrographic thin/polished section is a bit smaller than the dimensions of the carrier glass (minus an engraved or adhesive label), in practice near ~25 mm × 40 mm, for convenience rounded off to 1000 mm^2. The total volume of material comprised at 30 μm thickness (see below) amounts to 30 mm^3, relevant information in discussions about sample size and representativity (see preceding sections).

Regardless of outer dimensions, thin (or polished) section thickness is by default 30 μm, as on the widespread petrographic wall-chart of Lof [158] as well as in petrographic atlases (e.g., [57, 75] and references therein). This default thickness facilitates straight-forward comparison of interference colours of birefringent minerals from different literature resources. Also, 30 μm was a thickness that could be reliably achieved by a cunning operator, in the early days of petrography when thin sections were manufactured swirling a billet by hand on glass plates instead of simply holding it on a spinning cast-iron disc.

However, there may be good reasons to make sections thicker or thinner than the default 30 μm. To avoid too many overlaying mineral grains rendering positive identification complicated or impossible, ultra-thin sections are preferable for sample materials with grain sizes below say ~30 μm, also depending on the shape of grain contacts (e.g., straight, concave, serrate, interlobate). Typical examples include rhyolite, basalt, siltstone, limestone, chert/flint, granofels (formerly known as hornfels), and (damaged) concrete. Yet, there are practical limits to the achievable minimum 'thinness' of a thin section, determined by e.g., actual equipment used, specimen mounting and overall preparation procedure, as well as mechanical strength properties of the sample material, abrasion-hardness contrast between composing minerals, cleavage, presence of microcracks, porosity, degree of alteration, etc., which can only partly be compensated by operator skill. Using mechanical polishing, ultra-thin sections can be reduced to 10 μm. If even thinner sections would be needed, use of broad-ion-beam milling (BIB) equipment is required [68, 97].

Conversely, certain instrumental methods require sections thicker than 30 μm, notably laser-ablation (LA-) ICP-MS, in which sample material is ablated by a laser beam projected via an optical microscope onto the specimen surface in a vacuum chamber. Commonly, 'thin' sections for this purpose are finished at 100–500 μm, in addition to double polishing for laser ablation applications (to minimize scatter on the bottom face). Another ablative method is secondary-ion mass-spectrometry (SIMS) that uses an ion beam rather than a laser. A comparatively ordinary electron-probe micro-analyser (EPMA) instrument might also benefit from thick sections, especially when atomically light and texturally porous materials are to be assessed (e.g., ettringite), to keep the interaction-volume of the incident electron-beam within the sample material, in other words, to prevent the analytical result from being contaminated by signal from the underlying carrier glass.

In certain (rare) cases, requirements defined by the sample material may be in conflict with requirements defined by the analytical method. A compromise must be found, by adjusting section thickness (thinness), by adjusting instrument settings and operating conditions of the analytical method, or by choosing an altogether different method, by accepting analytical results of lower quality, or by a combination of these factors. A discussion between analyst and researcher will be helpful.

3.2.4.2 Media for Mounting and Impregnation, Use of Dyes

Nicol's and Sorby's original thin sections were hot-mounted to the carrier glass with Canada balsam, a tree resin of natural origin that fully melts at ~140 °C. For application as an adhesive, the resin is diluted with an organic solvent (e.g., xylene), which needs to be boiled off at ~180 °C to obtain full bonding strength; a rather delicate process [123]. Canada balsam remained the preferred mounting medium until WWII, after which it got quickly replaced by Lakeside 70 cement, sold in sticks for easy melting but not requiring cooking like Canada balsam does, and with superior adhesion to glass. However, many minerals have been recognized as sensitive to temperatures 'above ambient', arbitrarily say >65 °C. Well-known examples include ettringite (irrespective whether of natural origin or in hardened concrete), clay minerals, zeolite group minerals, etc., that all release water above ambient temperature. Thus, hot-mounting may attribute undesirable preparation artefacts to the specimen, and is therefore mostly abandoned. Of course, this applies unabridgedly to any other type of hot-mounting, including hot-curing epoxies. In the further, we will therefore refer to cold-curing epoxy types only.

Today, thin and polished sections are most commonly mounted using a two-component epoxy, consisting of a resin and a curing compound, latter also referred to as hardener or activator. As such, the term 'epoxy' is generic, and 'epoxy *sensu lato*' with a broad range of qualities in liquid as well as in fully-set state is available from many suppliers, hereunder thin sectioning/polishing equipment manufacturers. Instructions for mix proportions of components and application of the fluid mix are product-specific and are not given here.

For preparation of thin and/or polished sections, the ideal combination of resin and curing compound should have low viscosity (like thin olive oil), a high capillary affinity (i.e., effective wetting of the sample material, ensuring deep penetration), up to 1 h pot life while the mix still can be applied, a setting time of a few hours, low volatile content (to avoid boiling up under vacuum impregnation), suitable final hardness (comparable to the sample material), a refractive index close to quartz (n = ~1.545), no discolouration upon aging (browning over time, like e.g., Canada balsam and Lakeside 70 tend to do), and negligible shrinkage upon setting and aging.

Use of analytical instruments operating under (ultra-) high vacuum conditions (e.g., SEM–EDS, EPMA, LA-ICPMS, SIMS) is increasing, leading to additional quality requirements for the mounting medium, notably: suitable and consistent chemical composition (including ultralight and trace elements) not to compromise elemental analysis, low volatile content of the fully hardened epoxy to minimize outgassing in vacuo, low fluorescence so as not to interfere with CL and SE detectors, and low sensitivity for e-beam damage (only for SEM and EPMA). In-depth assessment of selected epoxies from different brands revealed substantial differences between products for all qualities mentioned above, including the non-vacuum requirements [236]. According to assay by SEM–EDS, in addition to the nominal carbon and oxygen (ideally), all products contain measurable amounts of chloride, most contain silicon and aluminium, a few contain sodium and potassium, whereas some contain lithium, boron, sulfur, chlorine, or yet others.

In summary, the choice for a given type of epoxy should be based on the particular combination of properties and qualities that best fit the particular job at hand, with the instruments available. An extensive review on additional mounting materials, methods, and equipment is given in Echlin [68].

Weak materials (e.g., friable, highly porous, finely crystalline, ductile or bendable, with low abrasion hardness, or perfect cleavage) are prone to mechanical damage by sawing, cutting, lapping, and polishing, attributing preparation artefacts that do not belong to the original material. Preparation damage is greatly ameliorated by impregnation, which reinforces and supports the material through immersion in mechanically robust and resilient epoxy. It should be noted that all qualities of epoxy will impregnate a sample to some extent simply by its wetting action, however, only a low-viscous epoxy of high capillarity is capable of fully infiltrating the finest details, and only an epoxy of sufficient hardness and rigidity is able to provide adequate mechanical support. Thus, complete impregnation ensures the preservation of the original microstructure, texture/fabric of a given material, and may even be a prerequisite for that (e.g., damaged concrete). In addition, pervading interstitial pore space with epoxy reduces chances for pull-outs, plucking, and undercutting (=topographic relief induced by contrasting abrasion hardness of adjacent phases), as well as it inhibits local specimen contamination by deposition of debris, from either liberated sample material or abrasive (see e.g., [134, 229]). The debris from pull-outs and plucking not rarely destroy a nearly finished or otherwise extra-precious section. In short, complete impregnation as described above can greatly save on time and effort, as well as annoyance.

Impregnation of pre-dried specimen is foretaken in vacuum (~0.01 mbar, or ~1 Pa), to remove water that otherwise would obstruct pores from filling with epoxy (epoxy being water-shy). Impregnated specimens are cured (and pre-dried) at 50 °C; higher temperatures will cure (dry) faster but might again introduce thermal artefacts, as discussed above.

To enhance visibility of pore space, sedimentologists commonly premix a blue dye (copper phthalo-cyanine complex [36]) to the resin, i.e., the most voluminous epoxy component to make sure it is completely dissolved [267]. However, the blue dye is easily overwhelmed at high magnifications when more light is needed, lamp power is increased, and a halogen bulb delivers more bluish light. This has mostly been replaced with admixing a fluorescent dye [35] that emits green light, close to where the human eye has its sensitivity maximum. The blue light component of an ordinary halogen bulb is well-sufficient to induce a green fluorescent emission, so expensive UV illumination (e.g., HBO - mercury arc, or XBO - xenon arc) is not necessary. In an ordinary petrographic microscope equipped with an inexpensive set of excitation and emission filters (available from all main instrument manufacturers), the fluorescent epoxy lights up green everywhere it has impregnated the sample material, whereas the background remains dark. Combined with rigorous preparation routine, fluorescence also enhances the identification of artefacts (see e.g., Danish Standard [64]). In certain cases, the presence of palagonite (altered volcanic glass) may prohibit use of Solvent Yellow 43 fluorescent dye [35], as its yellowish colour in plain light may be difficult to distinguish from palagonite. Additional details for fluorescence-impregnation of full cores and thin sections for deleterious AAR are given in Broekmans [26, 28].

In precisely plane-parallel sections, the fluorescent emission can be measured with a simple photo-cell (as used to measure reflectance of ore minerals in reflected light), or with a digital camera [69]. Each section is its own reference: excess fluorescent epoxy along the edges provides a 100% calibration point, and a completely non-impregnated impermeable area/particle presents the dark 0% point. Assuming section thickness is uniform and dye is fully dissolved in the epoxy, all intermediate values can be interpolated. By combining impregnated samples of concrete with different but precisely known water-to-cement ratios in a single section, observations on concrete samples with unknown water-to-cement ratios can be estimated with excellent accuracy and repeatability [266].

A sample briquette (impregnated or not) may be mounted on a carrier glass with a colourless epoxy (without dye), or another type of glue with a suitable combination of properties and qualities. A quick way to mount is using UV-setting acrylate glues, which achieve their full mechanical strength within 30 min, and have a very good adhesion towards glass necessary for further preparation. Protocol and materials for successful preparation of fluorescence-impregnated thin sections and full cores in geomaterials are elaborated in Broekmans et al. [28].

3.2.4.3 Particulate Materials and Block Mounts

Preparation of thin or polished sections from particulate sample materials calls for custom procedures. In a typical procedure, a subsample of the particulate material is mixed with the mounting epoxy in a polymer jar of ~40 mm width to a height ~25 mm, so that the cross-section is near the dimensions of a standard petrographic section. As the particulate-epoxy mix is prone to gravity segregation (heavier particles concentrate at the bottom), use of a wide and shallow jar is preferred over a narrow and tall one (see e.g., [52]). Overall representativity is safeguarded by the overall number of particles of all sections for a given size fraction together, not the number of points counted per section, as elaborated in Howarth and French ([119], pp. 174) and Broekmans and Fernandes [27]. When the epoxy has fully set, briquettes are sized with a diamond blade and thin sections prepared following regular routine. As the volume proportion of epoxy in thus prepared specimens is comparatively high, the final hardness of the epoxy becomes an important quality (see below), to avoid undercutting and pull-out of individual particles.

However, the main challenge remains avoiding segregation, caused simply by the effect of gravity on particles with different densities. Some interesting devices have been developed to avoid segregation of heavy ore minerals (e.g., [143]), to keep the particulate-epoxy mix rolling until it sets, so that segregation is avoided. However, this method requires use of custom equipment. Therefore, an alternative approach is to customize the properties of the mounting medium instead. Addition of small amounts of carbon-black nano-powder (2.2 wt% to the resin only) and mixing thoroughly for 3 h changed the consistency from normal flowing to thixotropic [19]. In terms of fluid-flow physics, the epoxy changed from normal to non-Newtonian-flow. Mixing in curing compound as usual and subsequent addition of the particulate sample matter completely prevented segregation. With simple rheological modelling using the Stokes law, it can be mathematically shown that the time needed for segregation in the modified epoxy is much longer than the setting time, meaning that it is effectively suppressed. The efficacy of modified epoxy by addition of carbon black has been tested for artificial binary 1:1 volume mixes of minerals with very different densities, i.e., graphite (2.1 $kg{\cdot}L^{-1}$), quartz (2.6 $kg{\cdot}L^{-1}$), pyrite (5.0 $kg{\cdot}L^{-1}$), and galena (7.4 $kg{\cdot}L^{-1}$), confirming no measurable segregation as well as excellent particle dispersion [19]. The overall preparation procedure and equipment remain unchanged.

One additional advantage is that the addition of carbon black renders the epoxy electrically conductive, so that a conductive coating for electron beam is not required. A disadvantage is that the addition of carbon black renders the epoxy opaque black, so that study of thin or polished sections in an optical microscope is no longer possible in transmitted light, only in reflected light (as for ore minerals). Therefore, ore minerals (concentrates) are usually cast in round epoxy 'blocks' Ø25 mm (~490 mm^2) or Ø30 mm (~710 mm^2) using widely available standard silicone rubber moulds, and polished. Thickness is for practical reasons commonly 5–10 mm. Apart from opacity, there is no technical reason against use of polished (carbon black modified) epoxy blocks for optical petrographic assessment of aggregate materials.

For electron probe beam instruments (SEM, EPMA), there is another important issue for discussion. The total volume of a polished block with dimensions as mentioned above lies between 490 $mm^2 \times 5$ mm = 2450 mm^3 and 710 $mm^2 \times$ 10 mm = 7100 mm^3. By comparison, a completely impregnated thin section of standard petrographic size contains maximum ~1000 $mm^2 \times 30$ μm = 30 mm^3 of epoxy plus an additional 5 mm^3 of acrylate to mount the specimen on the carrier glass, total 35 mm^3. In practice, the epoxy volume in the thin section is substantially reduced by the sample material present in the 30 mm^3 (but not of the 5 mm^3 acrylate mount), and of course the same applies to the polished blocks where the epoxy volume is reduced by the sample material present. Still, a polished block may thus contain up to 140–400 × the amount of epoxy compared to a standard thin section. It is widely known that ultrahigh vacuum instruments like SIMS need a long time (typically overnight) to pump down when fresh polished block specimens are placed in the specimen chamber, which is attributed to outgassing of the epoxy (also see [236]). This can be reduced by selecting a type of epoxy with low volatile content. Further reduction would be possible by removing 10 to 15 mm of material from the back of the specimen. That volume is not accessible for analysis anyway and can be replaced by a metal stub that does not outgas. For analysis in SEM or EPMA instruments, Ø25 mm blocks of reduced thickness can be affixed to a standard petrographic carrier glass using double-sided copper or carbon tape for easy mounting in the specimen holder.

3.2.4.4 Abrasives, Polishing

The original thin section of Sorby was ground to thickness by swirling a glass-mounted chip on a heavy glass plate with emery abrasive. Emery is natural rock-forming corundum α-Al_2O_3, a non-precious variety of sapphire used for grinding wheels and other abrasive applications, representing standard #9 on the Mohs hardness scale [173]. In many modern abrasive applications, emery has been replaced with synthetic silicon carbide (SiC, aka. carborundum, originally a trade name) which scores 9.5 on the Mohs scale, above sapphire but under diamond at 10. On the Mohs scale these differences appear only marginal, however, the entire scale is limited to only ten entries to cover the entire hardness span from talc (1) to diamond (10). Mohs designed his scale to be practicable for field use, not as a scientific instrument.

This shortcoming was recognized by Rosiwal who devised his own abrasion hardness scale about 70 y after Mohs [204]. The Rosiwal scale is based on internal standards quartz (set at 100) and corundum (emery, set at 1000) used as abrasives to determine relative hardness. A carefully weighed amount of abrasive is used to wear down an unknown mineral sample, until the abrasive is completely worn. The weight loss of the sample then is a relative measure for its abrasion hardness. Measured hardness for Mohs-scale minerals range from talc (0.03) to diamond (140000), a total span of ~4600000. Silicon carbide rates at 2000–5000, varying with quality of the synthetic product. For certain applications, a novel abrasive consisting of cubic boron nitride (cubic BN, abbreviated as CBN) with an even greater Rosiwal hardness of ~10000 is sometimes used.

The Rosiwal scale more accurately reflects mineral hardness differences than the Mohs scale is able to. It reveals the vastly superior hardness of diamond even compared to its next neighbours, explaining why diamond is the abrasive perfect. In addition, diamond is to some extent self-sharpening due to its octahedral cleavage, a property that corundum and silicon carbide both lack. Blunted particles in a silicon carbide slurry (a common way to supply abrasive during grinding and lapping) tend to start rolling between the specimen and the equipment. In contrast, a blunted diamond particle will chip off a thin flake exposing a fresh cutting edge, losing again its roundedness. Bonding diamond in a matrix of sintered metal eliminates the chance for abrasive to roll, and bonded diamond has become the preferred abrasive for preparation of thin sections (see e.g., [5]). The superior cut that diamond offers, especially if combined with impregnated thin sections, is unsurpassed by alternative abrasives.

Polishing represents the final stage of the mechanical part of thin section specimen preparation. Previous stages have consisted of sizing a specimen, mounting, and various stages of lapping to achieve the desired thickness. Each next stage with finer abrasive is to remove damage attributed by previous stages in the preparation protocol, e.g., surface roughness, scratches, undercutting, as elaborated above. Polishing of geomaterials is commonly done on cloth using diamond pastes of increasing fineness, typically 4 μm and 2 μm, and the final stage with 1 μm or even 0.25 μm. Both coarser and finer diamond paste varieties are available, but usually this type of polishing provides a mirror finish that is more than adequate for ore microscopy using incident polarized light, or for chemical spot-analysis by SEM-EDS or EPMA. However, in materials with great hardness contrast (e.g., chert in limestone, magnetite in serpentinite, damaged OPC concrete, bituminous concrete), undercutting of the soft phases often produces an undesirable artificial relief. In such cases, replacing soft cloth as a polishing medium with a harder polymer or metal disc may prevent undercutting. A wealth of products to fit every need are commercially available.

If a surface with an even higher polish and a still lower level of attributed damage is required (as for e.g., EBSD - electron backscatter diffraction, on SEM), the surface is polished with a colloidal suspension of silica or alumina, which have particle sizes down to a few tens of nanometres, i.e., roughly three orders of magnitude finer than diamond pastes. Together with a suspension medium of low pH, the polishing is of combined mechanical (by the suspended particles) and chemical (by the suspending medium) action.

3.2.4.5 Cover Slip or Cover Spray, Conductive Coating

One of the aspects of preparing a thin or polished section is preservation of the material in its present status, as mentioned above. The large nominal specimen area (~1000 mm^2) of a thin section of standard petrographic size combined with a low thickness (30 μm) renders the material accessible to external agents that may alter the material. Internal porosity of the sample material expands the exposed surface area beyond the nominal specimen area, whereas complete impregnation with epoxy (see above) may reduce it to below that value.

Alteration by exposure to air does happen; it is not just an invented phenomenon. Examples include carbonation along thin section edges by uptake of CO_2 from laboratory air or the discolouration of bluish-greenish GBFS in cement paste, both by (too) slow drying during preparation. Finished and finalized sections may develop efflorescence on sulfides from moisture-uptake from air, or develop a tarnish ('rainbow colours', or dulling/blackening) on sulfides by oxidation. Certain types of alteration are less or inconspicuous, e.g., evaporation of internal water in opal, analcime or other zeolites, exsiccation of clays or sulfates, or attraction of water by hygroscopic minerals. Some minerals are very sensitive and will visibly alter in a matter of days, hours, or even mere minutes; others appear not to be affected after decades. Alteration layers will also have a detrimental effect on microbeam analysis (e.g., [155]).

By far the most effective way to prevent post-preparation alteration is by mounting a cover slip, which seals off the specimen permanently. Cover slips are available from many suppliers in a range of sizes and formats, with thicknesses of 0.10–0.20 mm, in quartz or soda-lime glass, or polymer, even with a pre-etched reticle for correlative microscopy. Cover slips can be mounted with clear epoxy (without dye), with UV-curing acrylate, or alternative glue types. The most common objection against use of cover slips is that these never more can be removed, which however is not true. Certain types of 'permanent' polymer cover slips and glue are soluble in concentrated alcohol or another organic solvent and can be removed later in case of regret. Moreover, a cover slip can be mounted temporarily using paraffin oil or even plain water, though water might possible react with or dissolve certain minerals and is therefore not recommended. Furthermore, certain thin sectioning equipment allows very precise removal of a glue-mounted cover slip.

Mounting a glass or polymer cover slip greatly improves the quality of view in a petrographic (optical) microscope. It eliminates any remaining surface roughness on non-polished sections and thereby optical scatter, so that finer details are resolved rather than being blurred. An optical microscope is designed as an optical system, and the specimen is an integral part of the light path. Optimum viewing quality is achieved if the thin section is finished compliant with the actual instrument's optical design. On a modern petrographic microscope for thin section study in transmitted light, objectives will have engraved in red (for polarizing optics) below magnifying power and numerical aperture (e.g., "40×/0.85") an indication like "∞/0.17" (or "160/0.17" on older instruments with 160 mm optics), to signify the objective is designed for use with a cover glass (of approximately 0.17 mm thickness). A petrographic microscope for study of opaque ore minerals in incident light will also have objectives engraved in red, instead listing "∞/−" (or "160/−"), indicating the objective is designed for use without cover slip.

A popular alternative for a solid cover slip is a plastic cover from a spray can, which is quick and convenient to apply as well as easy to remove with an organic solvent (see [174]). However, optical improvement is only moderate as the surface is not quite as smooth and plane compared to glass (see [59, 175]). Applying a second layer of spray cover is technically possible, but at least halves speed, doubles labour, and of course adds complication. Given the optical and other advantages of a solid cover, spray cover is generally not recommendable.

Most rock-forming silicate minerals are electrical insulators. Therefore, polished rock/mineral sections intended for micro-analysis in an SEM or EPMA instrument must be coated with a conductive layer to minimize charging in the electron beam (see chapter 9 in Goldstein [97], 133ff). Even when the issue seems straight-forward, it is not quite, as argumented by Bastin and Heijligers [4], and reinstated by Echlin [68], pp. 248), the issue is still a topic of investigation [155]. Arguably the most common coating for SEM imaging is 'gold' or alloys with palladium (smaller grain size than pure gold), whereas for analytical work in SEM–EDS or EPMA the specimen is often coated with carbon. The choice coating is determined by the nature of the materials under investigation (here: 'geomaterials *sensu lato*'), the type of analysis, and the element species of interest (i.e., composition). The checklist in Echlin [68], pp. 259) counts eight entries. In principle any metal could be used, but some metals oxidize or tarnish quickly (e.g., silver) and are therefore less practicable.

For elemental analysis by EDS (or WDS) and/or analysis by cathodoluminescence (CL), it is important that a coating negligibly attenuates emitted radiation (soft X-rays and Auger, characteristic X-rays, light) for detection. Absorption is determined by the atomic mass (Z-number) of the coating material, so an atomically light material (e.g., carbon, atomic mass ~12) absorbs less than a heavy material (e.g., gold, atomic mass ~197). Also, a thin coating absorbs less than a thick. Kerrick et al. [133] provide a simple means to evaluate carbon coating thickness using a color scale (Table 2, p. 924), listing thicknesses of 150 Å up to 450 Å in 50 Å increments (15–45 nm in 5 nm increments).

Coatings are applied under vacuum either by plasma sputtering, or by thermal evaporation; see Echlin [68] for further detail. A broad choice of coating equipment is available from many suppliers in 'entry level' to 'highly advanced' varieties. To achieve an even coating of uniform thickness, some equipment places specimen on a planetary movement stage during the coating process. For elemental analysis by SEM–EDS or EPMA, a coating thickness of 100–200 Å (10–20 nm) is common, but for high resolution imaging of nanometre-sized features, coatings must be accordingly thinner to avoid obstructing visual detail. Coating thickness is effectively monitored to 0.1 nm accuracy using a piezo-electrical quartz crystal. The deposited coating material changes the crystal's mass and therefore its vibration eigen-frequency, which can be measured with very high precision.

Subsequent instrumental analysis may require removal of an existing coating. Here, impregnated sections have the advantage that their surface is 'closed', so that a coating cannot be deposited in voids or cracks (as they are all filled with epoxy already). Carbon coating may be removed by gently rubbing with lint-free tissue or microfibre cloth moisturized with isopropanol (*pro analysi* grade). Alternatively, some coating devices have plasma etching options, which are very effective. Acetone is not recommended, as epoxy and acrylate cements may decompose or dissolve. For carbon coatings that prove resilient to mere rubbing, and for all metal coatings, careful manual polishing using diamond on cloth is needed. Gold coating may be chemically dissolved [154], but compatibility of the mounting epoxy must be verified first. Use of a metal specimen holder (see e.g., [47]) is recommended, to ensure original section planarity is maintained.

3.2.4.6 Maintenance, Travel and Storage

Fingerprints and many other accidental stains can be removed from a section surface (polished or with a cover slip) with lens tissue wipes, i.e., very soft lint-free paper specially made for wiping microscope objectives, oculars, and other optics parts. The surface of polished sections with a conductive carbon or metal coating must not be touched to avoid damaging the coating. Instead, they might be attempted cleaned by blowing oil-free compressed air or dry nitrogen to remove dust particles. If the instrument has a plasma cleaner installed (occasionally on the load lock), then persistent stains may be removed inside the vacuum chamber. Consult the operator for information.

Ready prepared thin, polished, covered and/or coated sections can be kept in boxes from inexpensive folded cardboard or more durable wood, hosting 10–200 standard sized petrographic specimen. Secure sections against movement during transport and travel with a piece of thin foam plastic in the lid, or a few loosely rolled sheets of paper kitchen towel. Clearly mark top–bottom on the outside of the box as well as content (e.g., "careful - microscopy specimen"), to minimize chances for sections flying around if opened by a security officer unfamiliar with such specimens.

For long-term storage, prepared sections are to be kept in the dark, in a conditioned space at (constant) cool room temperature and normal indoor humidity (RH ~30–50%). Though modern media are little sensitive, cool-dark-dry conditions will prevent browning of the mounting medium, as well as photo-bleaching of a dye if present. Latter is especially important for calibrated thin sections as mentioned above [266].

3.2.5 Analysis of Alkalis of Aggregates by Electron Microscopy and Microanalysis Techniques

The alkalis content of the aggregate is usually determined by chemical analysis of bulk whole-rock, after grinding the rock or sand to powder. It is important, however, to study the mineral composition of the aggregate, as the solubility varies with the type of mineral that exist in the rock. According to the methodology for the assessment of potential alkali reactivity of an aggregate, petrographic characterization of the rock types should be the first step, followed by expansion laboratory tests ([80, 215], RILEM AAR-7.1, 2016; [75]). A visual assessment of aggregates can be obtained swiftly (macroscopic characterization at hand sample scale) by the identification of the rock type (from quarries) or the polymictic aggregate components (in unconsolidated sand and gravel of mixed composition). This is usually performed by manual separation of the particles according to colour, texture and features visible by unaided eye with the help of a hand lense and, if available, a stereo microscope.

However, the identification of the minerals that constitute the aggregate, their modal contents, as well as structure, texture/fabric, porosity and degree of alteration demand microscopic examination of thin sections, according to Humphries [123] and Fernandes et al. [75] (Fig. 3.5). The texture is of utmost importance as it defines the facility of pore fluids from the concrete to penetrate the aggregate and enhance the dissolution of the minerals. The degree of alteration evaluated in thin section presents an indication for the potential of alkalis release from the aggregates (Fig. 3.6). It has been confirmed in the study of granitic samples developed by Ferraz et al. [77] that the weathered rock will release less alkalis than the same one in a sound state (non-weathered). The content of each mineral, whether sound or altered can be done by visual estimation or (preferably) by point counting.

Optical thin section petrography may be complemented by scanning electron microscopy (SEM) with in-situ chemical analysis (EDS, Energy Dispersive Spectroscopy), which is essential when studying very fine-grained rocks (Fig. 3.7), or supplemented with quantitative chemical analysis by Electron Probe Micro-Analyser (EPMA). Combined with the chemical analysis of bulk rock, petrographic examination is essential in order to verify the presence of reactive forms of silica present in the aggregate but also the type of alkali-rich minerals that might contribute to the development of the ASR.

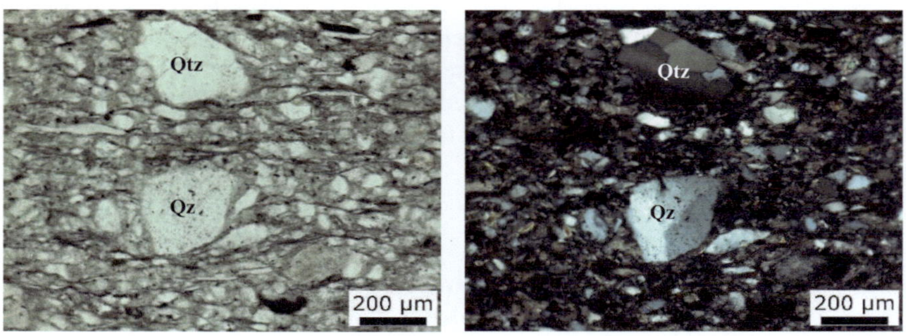

Fig. 3.5 Greywacke aggregate in thin section: the mineral content and texture can be studied in detail in an optical microscope with different illuminations (left: plane-polarized light; right: cross-polarized light). This is a meta-sedimentary rock with coarse clasts of quartz (Qz) and quartzite (Qtz) and a finer matrix composed of microcrystalline mica, quartz, clay minerals and organic matter, defining a preferred orientation of the minerals due to burial compaction and diagenesis

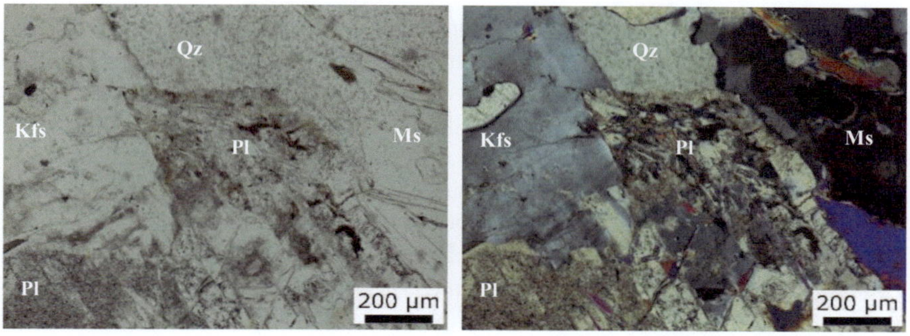

Fig. 3.6 Example of igneous rock with weathering, which is visible in the cloudy appearance of the K-feldspar (Kfs) and plagioclase (Pl, Na-Ca) grains. Image on the left is in plane polarized light, on the right in crossed polarized light. Ms, muscovite; Qz, quartz

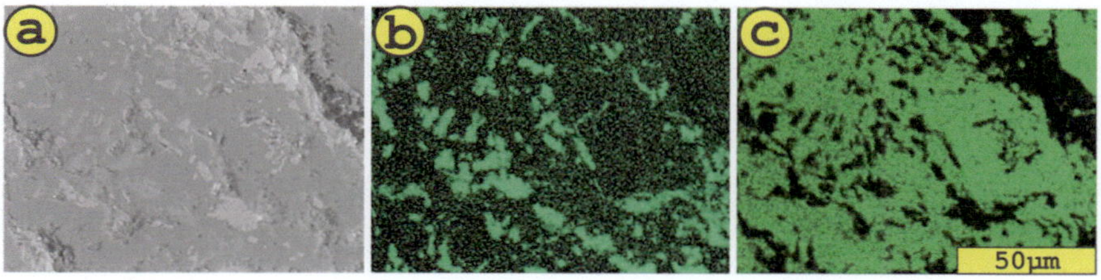

Fig. 3.7 SEM micrograph from fine-grained cataclasite in **a** backscattered-electron (BSE) imaging mode, **b** EDS-element map for K revealing presence of K feldspar, and **c** for Na revealing the presence of albite (=Na-plagioclase). Edited from Menéndez et al. [169]

3.2.6 Extraction of Pore Fluid from Hardened Concrete

The presence of sufficient alkalis (Na^+, K^+) in the concrete pore solution increases the pore solution hydroxyl (OH^-) concentration above a certain threshold level to dissolve silica in the aggregates. This threshold hydroxyl concentration varies depending on the type of aggregate lithology, but is often in the range of 0.20 to 0.25M [58, 140]. The cement is typically considered as the primary source of alkalis, but certain SCMs (particularly some high-lime fly ashes) and aggregates may also act as internal sources of alkalis over the life of the structure, while deicing salts and seawater have the potential to act as external alkali sources [194].

Determination of the alkali and hydroxyl content in pore solution can help engineers assessing an ASR-affected structure make a prognosis regarding the potential for future expansion, particularly if multiple measurements are made to monitor changes over time. ASR gel formation consumes alkalis from the pore solution, leading to a reduction in the pore solution alkalinity. Concrete with a highly-alkaline pore solution is more likely to experience continued expansion from ASR, while a less-alkaline pore solution is unlikely to continue dissolving silica. In some cases, alkali recycling (exchange of Na and K in the ASR gel for Ca) can cause the pore solution alkalinity to stabilize above the threshold level and potentially allow ASR to continue so long as reactive silica remains available [107, 139, 238]. Concrete with aggregates or SCMs releasing alkalis to the pore solution over time may see a net increase in the available alkali content as ASR progresses [66, 87, 95].

There are different methods for measuring the available alkali content of hardened concrete. These techniques all require laboratory processing and testing of samples taken in the field. There are no known techniques for measuring alkalis in the field. It should be noted that alkali contents may vary throughout the structure, sometimes quite widely [212], particularly if multiple sources of materials were used for construction.

Some techniques such as the pore pressing are carried out directly on a concrete sample whereas other methods are carried out on finely crushed concrete samples. A recent review [191] details the different methods used to measure the pH but also the free alkali content of the concrete pore solution.

3.2.6.1 Pore Pressing

The most commonly used method to characterize the chemistry of the concrete pore solution is the high-pressure extraction method, also known as pore pressing. This method was originally proposed by Longuet et al. [160] and by Barneyback and Diamond [3]. The pore pressing procedure consists in placing a sample of concrete, mortar or paste in a specially-designed steel die on which high pressure of a few hundred up to 1000 MPa is applied to recover the extracted interstitial fluid. The small volume of expressed solution (on the order of a few mL for 200–300 g of concrete) is then diluted and analysed to determine its chemical composition. Barneyback and Diamond [3] contains the most complete description of the analytical methods for each ionic species found in the pore solution, although any method suitable for aqueous solutions may be used. Hydroxyl concentration (OH^-), and therefore pH, are typically determined through titration against HCl; the use of glass electrode pH meters is not recommended because of the aggressive nature of the solution. The analysis of the chemical composition of this liquid is carried out after acidification by atomic absorption spectrometry (AAS) or inductively coupled plasma atomic emission spectroscopy (ICP-AES). Hydroxyl concentrations are also sometimes simply assumed to be in balance with the sum of the sodium and potassium concentrations; in the case of mature concretes where other ionic species have decreased to comparatively insignificant levels, a widely accepted simplification [14, 65, 131]. If the concrete is very old or when the water-to-binder ratio is low, the amount of fluid extracted during the test may be insufficient to perform the chemical analysis.

In general, the volume of fluid extracted by this method represents only a small fraction of the total pore solution content of the concrete. This raises questions about representativity of the chemistry of the fluid extracted by this method compared to the concrete pore solution. An increase of pressure could have an influence on the pore solution composition. Duchesne and Bérubé [65] extracted pore solutions from cement pastes (water-to-binder ratio of 0.5) with various SCMs under pressure intervals of 0 to 560, 0 to 200 and 200 to 560 MPa. Authors showed that alkali concentrations were not affected by the pressure at which samples were expressed. Similar results were reported by Tritthart [242] for pressure up to 3500 bar (=350 MPa). However, Bérubé and Tremblay [14] observed that the chemistry of the extracted solution varies according to the pressure applied for very high pressures up to 1000 MPa (=10 kbar). This last fact is supported by Chappex and Scrivener [43] with applied pressure up to 800 MPa. The effect of the extraction pressure on the composition of the pore solution shows contradictory results. It is possible that these variations are due to peripheral influences associated with the loading rate, the increase of the temperature at high pressure or the condition of the sample tested (monolithic or crushed samples). In fact, Tritthart [242] concluded that breaking up of samples before pore pressing should be avoided due to contact with air containing CO_2. Further research is needed to clarify this point.

Recently, Drolet et al. [62] evaluated the variability of the extraction method on six mortar and six concrete samples and obtained coefficients of variation of 1.56% and 1.21%, respectively. This shows that the extraction technique is reproducible and represents the composition of the pore solution if reasonable pressures are used during the test. Vollpracht et al. [254] state that a pressure of 250 MPa (=2.5 kbar) should be sufficient for collection of pore solution from cement pastes.

The concentration of alkali and hydroxide ions in the pore solution of a cement paste is closely related to the initial alkali content of the cementitious material used [228]. Vollpracht et al. [254] reviewed a large number of experiments to analyse the chemistry of the pore solution of blended cement pastes. These authors reported data for cement pastes made with water-to-cement ratio of 0.5. In that case, the alkali concentration of the pore solution for a cement with a low alkali content (0.20–0.31 wt% Na_2O, 0.48–0.75 wt% K_2O) varies between 100 and 200 mmol Na^+/L and between 400 and 450 mmol K^+/L, whereas for a cement with a high alkali content (0.32–0.43 wt% Na_2O, 1.04–1.30 wt% K_2O), the concentration of the pore solution varies between 250 and 350 mmol Na^+/L and between 400 and 800 mmol K^+/L.

3.2.6.2 Hot Water Extraction Method

Water-soluble alkalis in concrete can be measured by allowing the alkalis (and other water-soluble components) to leach from a powdered concrete sample in water. In this method, the alkali content of the pore solution is extracted from powdered concrete sample immersed in water. The immersion solution is filtered and its alkali content is analysed. The goal is to extract only the alkali content available, or active, in the pore solution, but avoid extracting those bound in hydration products or ASR gel. Whether this is actually the case is a matter of debate. Davies and Oberholster [54] suggested that the test may extract alkalis from amorphous (but not microcrystalline) ASR gels. Diamond [58] suggested that some alkalis in the pore solution would not be recovered as a result of the concrete being dried before the extraction process.

The hot-water extraction method described by Rogers and Hooton [202] and Bérubé et al. [13] is a variation on the water-soluble alkali test found in ASTM C114. The technique involves boiling a 10 g powdered concrete sample in 100 mL of deionized water for 10 min, and then allowing the sample and water to rest at room temperature for 24 h. The sample solution is then filtered and analysed for its sodium and potassium content by atomic absorption photometry or inductively-coupled plasma spectroscopy. An advantage of this method over the pore solution expression technique is that a specialized die and high-capacity loading frame are not required.

An additional challenge with using this technique is that some aggregates might release alkalis during the extraction process. If these are not accounted for, then the available alkali content will be overestimated. In particular, aggregates containing alkali feldspar minerals can be expected to contribute additional alkalis and bias test results [12, 49, 186]. The potential contribution by aggregates can be determined by subjecting a sample of the same aggregate (if available) to the hot-water extraction process, and adjusting the measured alkalis by the weight fraction of that aggregate in the concrete.

The hot water extraction method, first described by Rogers and Hooton [202, 203]), is currently used in Canada to determine the water-soluble or active alkali content of field concretes. The latter is performed for determining whether or not the available alkali content is sufficient to maintain ASR. In the method proposed by Rogers and Hooton [203], mortar bars were broken, crushed and pulverized to pass a 75 μm sieve. One 1 g of powder sample was added to 100 mL of distilled water, boiled for 10 min, and allowed to stand overnight. The suspension was then filtered and the resulting solution made up to 100 mL with distilled water. The alkalis, Na^+ and K^+, were determined by flame photometry.

Bérubé et al. [9] proposed changes to the original procedure to improve the precision and the reliability of the hot-water extraction procedure and to simplify the experimental procedure. First, the concrete under study is dried under vacuum at a temperature of 80 °C until a constant weight is obtained, then crushed so that the entire sample is reduced to pass a 150 μm sieve. This choice was motivated by the fact that the step is easier to achieve and limit the alkali contribution by aggregates during the test. Two 10 g subsamples of the concrete sample under study and two 10 g subsample of a pulverized control concrete with a well-known water-soluble alkali content are immersed in 100 mL of distilled water, boiled for 10 min, and allowed to stand overnight at room temperature. Then, the suspensions are filtered and their volumes are brought to 100 mL by addition of distilled water and their Na^+ and K^+ concentrations were analysed. The average results for the two subsamples are normalized by comparison with respect to the value obtained for the control concrete. This correction, based on normalization from the value obtained on a control sample, reduces the coefficient of variation from about 20% to about 10% [9].

3.2.6.3 Hot-Water Extraction: "Espresso Method"

This method is a modified version of the hot water extraction method discussed above, recently developed at Laval University by Fournier et al. with the aim of improving repeatability and representativity [82]. In fact, it was difficult to control the boiling intensity and to recover all the material during filtration in the original method. To remedy these problems, the Espresso version has been proposed.

The Espresso extraction method first involved the crushing, grinding and pulverizing of a sample of mortar or concrete specimen to pass a 150 μm sieve. Then, a 10 g subsample is placed in a Büchner funnel with a filter paper. Then, 300 mL of boiling distilled water is filtered under vacuum through the pulverized subsample. The solution thus obtained is transferred into a 500 mL volumetric flask and allowed to cool down before the volume is completed to the gauge with distilled water. The alkali content is then determined by appropriate instrumental analytical methods. The Espresso variation of the hot-water extraction method is less time consuming and shows a better repeatability compared to the regular method described in the preceding section.

The alkali concentration in the pore solution can be calculated knowing the water-soluble alkali content and the water content of the concrete [12]. According to these authors, the measured water-soluble alkali content must be corrected to consider the alkali contribution by the aggregates during the extraction. In addition, these authors reported that the experimental variability of the hot-water extraction method remains too important to follow the progress of ASR in each concrete structure. Moreover, the results obtained indicate that significant amounts of alkalis incorporated in ASR reaction products are released in the test procedure.

The main criticisms associated with the hot-water extraction method concern the effect of drying of the samples prior to the extraction. Indeed, according to Constantiner and Diamond [48] and Rivard et al. [200] the alkali content of the concrete pore solution after drying and rewetting of the concrete sample is lower than for the same concrete that has not been subjected to drying. Nevertheless, Bérubé and Tremblay [14] compared results obtained by the hot-water extraction and the high-pressure extraction method. For a cement paste made with a water-to-binder ratio of 0.5, the measured alkali-equivalent concentrations of the pore solution were similar within 2% for the two extraction methods. In contrast, Drolet et al. [62] presented results in an attempt to determine the influence of aggregates on the concrete pore solution composition on the development of ASR. The expansion and the alkali content of the pore solution was monitored over time for a number of mortar and concrete specimens made with two types of reactive aggregates with different alkali contents and one non-reactive, alkali-free aggregate. Authors concluded that the hot-water extraction method overestimates the amount of available alkalis in the pore solution and generates a greater variability in the case of concrete specimens compared to results obtained by the pore pressing method. The results suggest that the hot-water extraction method released a part of the alkalis bound in cement hydrates. Moreover, a part of the alkalis fixed in the ASR secondary products seems to remain fixed during the leaching test.

3.2.7 Extraction of Alkalis from Aggregate Materials

The solubility of aggregates depends on several factors, namely the types of alkali-bearing minerals, the proportion of these minerals in the aggregate, the mineral grain size as well as aggregate particle size, the ratio of pore solution/aggregate, the pH and the type of alkali ions in pore solution [257, 258].

The study of releasable alkalis from aggregates has been performed using distilled water, saturated solutions such as lime, calcium hydroxide and alkali hydroxide solutions (e.g., [11, 13, 15, 49, 108, 132, 163, 167, 244, 245]) has analysed the different parameters used in the tests for the assessment the potential alkali release form aggregates and their contribution to ASR.

Since the first results of Hansen [108], many studies have been carried out to characterize the release of alkali by aggregates in the pore solution of concrete. Some authors have chosen to immerse aggregates in different types of solutions while others have worked directly on concrete or mortar samples by measuring the influence of aggregates on the pore solution extracted under high pressure. Different methods are described below and grouped according to the extraction method or medium chosen.

3.2.7.1 The "Espresso Method" for Aggregates

- Hansen [108]
  The first test to evaluate the alkali release by concrete aggregates was reported by Hansen [108]. This author has made mortar bars using sands normally used as a fine fraction in concrete. The materials used include sand from the Kaw River in Kansas (3.24 wt% Na_2O_{eq}), another from Old Lytle Creek, California (5.19 wt% Na_2O_{eq}), a high alkali cement (0.91 wt% Na_2O_{eq}) and a low alkali cement (0.20 wt% Na_2O_{eq}). The mortar bars were immersed in 200 mL of distilled water in a sealed container of 400 mL left at room temperature. Then, the Na_2O and K_2O content of the water was periodically analysed by flame atomic absorption.
  In presence of the Kansas sand, the alkali release in water after one year was 0.815 wt% and 0.18 wt% for samples made with the high and low alkali cements, respectively. These values represent 89% and 90% of the total alkali content of the virgin cements. It is therefore not possible to evaluate the alkali contribution of this sand. In presence of the Californian sand, however, the alkalis leached in distilled water after one year were 0.911 wt% and 0.262 wt%, representing 100.11% and 131% of the initial equivalent alkali content of the high and low alkali cements, respectively. The excess in the alkali content was attributed to a sand contribution.

- French [83]
  Different materials, natural sand, weathered gabbro and andesite, were studied by French [83]. The experimental protocol selected consisted of immersing 5 g of

material with particle size ranging from 0.05 to 0.15 mm in two 100 mL solutions, distilled water, and a solution with 10% by volume of HNO_3, stored at 23 °C and stirred for 3 days. The alkali content measured ranged from 0.008 wt% to 0.021 wt% Na_2O_{eq} in water and from 0.027 wt% to 0.066 wt% Na_2O_{eq} in the HNO_3 solution. It is important to note that these results were obtained from an acid attack, and are therefore not necessarily representative of the concrete pore solution.

3.2.7.2 Alkaline-Solution Extraction

Extraction in Calcium Hydroxide Solution

- Van Aardt and Visser [244–246]
  Six feldspars and two clay minerals were soaked in calcium hydroxide solutions. The procedure consisted of immersing a 0.5 g powder sample in a polyethylene container with 25 mL of water and 0.5 g of $Ca(OH)_2$ (water saturated with excess of calcium hydroxide). The containers were sealed and stored at a temperature of 39 °C for a period of 250 days. Results showed that feldspar minerals are prone to release alkalis in the lime solution, thus supporting the hypothesis that these minerals can potentially increase the alkali concentration of the concrete pore solution. The values obtained during this experiment varied between 0.02 wt% and 2.38 wt% Na_2O_{eq}. The maximum values reached during this experiment seem abnormally high, which can be explained by the small particle size (powder) used during the experiment.

- Stark and Bhatty [224]
  Ten granular materials were leached in $Ca(OH)_2$ solution to measure their alkali released. Different series of tests were carried out under different conditions. First, 5 g of material crushed and pulverized to pass a 74 μm sieve (No. 200) was placed and sealed in polypropylene bottles containing 25 mL of saturated calcium hydroxide solution. A first test series was conducted at 38 °C for periods of 7, 28 and 90 days. The samples were continuously stirred to prevent the precipitation of secondary materials. A second test series was sealed for 28 days in stainless steel containers, immersed in water at a constant temperature of 80 °C. A parallel series using distilled water instead of the calcium hydroxide solution was also sealed under the same conditions. These last two series were not agitated. At the due date, the solutions were filtered and analysed by atomic absorption to determine their Na^+ and K^+ content. After 90 days at 38 °C, the alkali content measured varied between 0.26 wt% and 1.99 wt% Na_2O_{eq}. At 80 °C, values between 0.82 wt% and 5.03 wt% Na_2O_{eq} were obtained in the saturated calcium hydroxide solution and between 0.03 wt% and 0.13 wt% Na_2O_{eq} for leaching in distilled water. These results seem abnormally high and may be due to the very fine particle size of the material tested.

- Le Roux et al. [153]
  In this study, the only standardized experimental protocol for extracting alkalis from aggregate particles was followed. This standard procedure is the LCPC No. 37 procedure [151], which comprises immersion of 500 g of sample particles with a size distribution between 0.08 and 20 mm, in 500 mL of a lime saturated solution with excess lime at a temperature of 100 °C for a period of 7 h without stirring. One of the advantages of this experiment is that it uses the same particle size as that normally used when making conventional concrete.
  The results obtained by Le Roux et al. [153] are shown in Table 3.3. The soluble alkali contents measured ranged from 0.0014 wt% to 0.21 wt% Na_2O_{eq} (mean 0.051 wt%); the highest value was obtained with the basalt sample, while the lowest was obtained by the high-purity limestone. Although basalt was not the material with the highest total alkali content, it was still the one that releases the most. The authors explained this by the low degree of crystallinity of the material. In general, vitreous materials as well as those containing a lot of feldspar minerals are more prone to release alkalis than siliceous and carbonate aggregates.

- Kawamura et al. [132]
  The contribution of aggregates containing alkalis to the development of ASR was investigated by Kawamura et al. [132] who analysed two highly reactive andesitic rocks containing feldspars in a calcium hydroxide solution. They found that the

Table 3.3 Alkali release as Na_2O_{eq} (in wt%) from aggregate upon immersion in saturated lime solution [153], or in alkali solution [185]. Pure limestone has very low content in alkali-containing minerals. Total alkali content in kg·m^{-3} calculated for concrete containing 1 850 kg m^{-3} of said aggregate. Adapted from Bérubé and Fournier [15]

Aggregate lithology	Immersion in saturated lime solution [153]				Immersion in alkaline solution [185]			
	Total Na_2O_{eq} (wt%)	Soluble alkali content			Total Na_2O_{eq} (wt%)	Soluble alkali content		
		wt% aggregate	wt% total	in kg·m^{-3}		wt% aggregate	wt% total	in kg·m^{-3}
Basalt	4.54	0.21	4.8	3.89	4.46	0.47	10.5	8.70
Granite	6.52	0.02	0.30	0.37	7.0	0.038	0.54	0.70
Rhyolite	5.50	0.036	0.64	0.67	5.92	0.041	0.69	0.76
Granitic gneiss	7.02	0.20	1.42	1.85	6.98	0.058	0.83	1.07
Amphibolite	3.67	0.02	0.5	0.37	3.37	0.03	0.89	0.56
Siliceous limestone	0.42	0.009	2.1	0.17	0.84	0.014	1.6	0.26
Quartzite	0.35	0.0085	1.8	0.16	0.90	0.008	0.8	0.15
Average	4.01	0.05	1.92	0.94	4.21	0.09	2.26	1.74
Pure limestone	0.036	0.0014	3.8	0.03	–	–	–	–

OH^- concentration was much lower than when just quartz was used as aggregate. In the test, 100 g of aggregates with a granular particle size distribution ranging from 0.15 to 5 mm similar to the fine fraction normally used in concretes or mortars was immersed in 200 mL of saturated lime solution kept at 40 °C, for a period of 180 days. Values of 0.02 wt% and 0.07 wt% Na_2O_{eq} were measured as soluble alkali contents for both aggregates.

- Lu et al. [163]
  Lu et al. [163] studied the granitic aggregate used in the Three-Gorges concrete dam in China and compared with the results from a gneiss and a feldspar by extraction in distilled water and super-saturated calcium hydroxide solution. The total alkali contents of the aggregates studied were 6.19 wt%, 3.87 wt% and 7.65 wt% Na_2O_{eq} for granite, gneiss and feldspar, respectively. In the test, 25 g of material were immersed in 50 mL of a saturated calcium hydroxide solution with 1 g of excess solid lime in sealed bottles placed in a water bath maintained at 80 °C. After 3, 7, 14, 21 and 28 days, the bottles were stirred and then 10 mL of solution were taken to perform the chemical analysis by atomic absorption. The different particle size fractions tested were 10–12, 5–10, 1.25–5, 0.63–1.25, 0.15–0.63 and 0.08–0.15 mm. The extraction procedure was also repeated with distilled water for comparison. The results of the study are presented in Table 3.4. Gneiss, containing micas and altered feldspar, released more alkalis than granite with unaltered feldspar. The researchers also observed that the finer the particles and the higher the temperature, the greater and faster the alkali is released. Finally, this study has shown a linear relationship between the alkali release by aggregates and the particle size for particle size ranging from 1 to 11 mm, while a drastic increase was observable for finer fractions.

Extraction in Alkali Hydroxide Solution

- Bérubé et al. [10, 12]
  Bérubé et al. [12] tested 17 aggregate types from Canada that present different alkali contents and various degrees of alkali reactivity. For each aggregate, 40 g subsamples of 1.25–5 mm particles were immersed in 40 mL of distilled water (K and Na supplies), lime-saturated solution with an excess of 1 g of solid lime (K and Na supplies), 0.7N NaOH (K supply), and 0.7N KOH (Na supply), always agitated and stored at 38 °C. After 7, 28, 112, 335, and 578 days, the samples were allowed to rest for 24 h at room temperature and a 10 mL sample of solution was taken out, and replaced by the same amount of the corresponding original solution. In each series of analyses, the four stock solutions were also analysed. The cumulative amounts of K^+ and Na^+ released by the aggregates with time were corrected for the alkali content in the stock solutions as well as for the amounts removed from the containers when sampling the solutions for the periodic measurements. The results of the study are presented in Table 3.5.

Table 3.4 Alkali released from different size fractions for selected aggregate lithologies, immersed in different solutions at 80 °C, as a percentage of the total Na_2O_{eq} (in wt%) in aggregate. Adapted from Lu et al. [163]

Aggregate lithology	Size fraction (mm)	Immersion in water, duration (days)					Immersion in over saturated lime solution, duration (days)					
		3	7	14	21	28	3	7	14	21	28	56
Granite	10–12	0.018	0.023	0.025	0.028	0.033	0.026	0.037	0.041	0.047	0.059	0.097
	5–10	0.021	0.026	0.029	0.032	0.037	0.029	0.041	0.049	0.055	0.067	0.120
	1.25–5	0.028	0.035	0.039	0.046	0.051	0.040	0.055	0.064	0.079	0.091	0.170
	0.63–1.25	0.039	0.045	0.056	0.063	0.082	0.060	0.072	0.084	0.092	0.126	0.237
	0.15–0.63	0.064	0.074	0.081	0.101	0.107	0.107	0.118	0.133	0.154	0.185	0.359
	0.08–0.15	0.128	0.143	0.153	0.177	0.191	0.232	0.254	0.307	0.370	0.457	0.829
Gneiss	10–12	0.033	0.043	0.059	0.071	0.084	0.070	0.091	0.126	0.144	0.225	0.423
	5–10	0.041	0.069	0.101	0.128	0.200	0.091	0.148	0.189	0.262	0.383	0.763
	1.25–5	0.069	0.137	0.192	0.250	0.304	0.217	0.314	0.368	0.450	0.529	0.955
	0.63–1.25	0.118	0.184	0.266	0.334	0.382	0.313	0.440	0.486	0.560	0.735	1.192
	0.15–0.63	0.168	0.302	0.369	0.404	0.567	0.438	0.572	0.647	0.747	0.989	1.648
	0.08–0.15	0.386	0.461	0.557	0.643	0.897	0.874	1.040	1.112	1.322	1.663	2.889
K-feldspar	10–12	0.010	0.012	0.014	0.016	0.018	0.043	0.056	0.067	0.079	0.091	0.165
	5–10	0.011	0.014	0.016	0.018	0.022	0.054	0.064	0.077	0.093	0.112	0.198
	1.25–5	0.012	0.016	0.021	0.025	0.028	0.061	0.078	0.098	0.114	0.134	0.252
	0.63–1.25	0.014	0.020	0.025	0.029	0.035	0.079	0.104	0.128	0.148	0.191	0.353
	0.15–0.63	0.022	0.028	0.032	0.036	0.048	0.108	0.138	0.161	0.195	0.246	0.417
	0.08–0.15	0.046	0.061	0.072	0.083	0.095	0.250	0.302	0.370	0.424	0.519	0.981

Table 3.5 Total alkali contents and soluble alkali contents for selected aggregate lithologies, in wt%. Column MAX gives maximum amount of Na_2O plus maximum amount of K_2O released, presented as Na_2O_{eq}, relative to the Na_2O_{eq} in aggregate, in percent (wt%). Prefix ">" indicates analysis was terminated after 4 months (tuff, Sudbury, Potsdam), or 11 months (greywacke, quartzite) due to precipitation upon acidification for instrumental analysis. Adapted from Bérubé et al. [10]

	Total alkali content (wt%)			Soluble alkali content from the aggregate (wt%)									
				Distilled water			Lime oversaturated			Alkaline solution (0.7N)			MAX
Aggregate lithology	Na_2O	K_2O	Na_2O_{eq}	Na_2O	K_2O	Na_2O_{eq}	Na_2O	K_2O	Na_2O_{eq}	Na_2O	K_2O	Na_2O_{eq}	wt%/ total
Phonolite	9.09	4.88	12.30	0.130	0.031	0.151	0.100	0.015	0.110	0.560	0.189	0.684	5.56
Granite	3.50	4.83	6.68	0.024	0.045	0.053	0.026	0.084	0.081	0.059	0.192	0.185	2.77
Rhyolitic tuff	5.25	1.70	6.37	0.053	0.010	0.059	0.065	0.018	0.077	>0.053	>0.022	>0.067	>1.24
Granitic sand	2.85	3.65	5.25	0.013	0.022	0.028	0.019	0.016	0.03	0.010	0.031	0.030	0.75
Diorite	4.49	1.13	5.23	0.028	0.006	0.032	0.038	0.023	0.053	0.136	0.065	0.179	3.42
Granitic gneiss	3.57	2.01	4.89	0.021	0.022	0.036	0.027	0.038	0.052	0.079	0.095	0.139	2.84
Andesite	4.47	0.12	4.55	0.047	0.001	0.048	0.056	0.001	0.057	0.101	0.002	0.102	2.25
Lithic gravel[a]	3.01	2.03	4.35	0.033	0.006	0.036	0.042	0.014	0.052	0.146	>0.032	>0.166	>3.83
Greywacke	2.33	2.60	4.04	>0.026	0.015	>0.036	0.025	0.012	0.033	0.108	>0.055	>0.145	>3.58
Shale	0.98	3.41	3.22	0.044	0.026	0.061	0.036	0.022	0.050	0.034	0.020	0.047	1.88
Dolostone[b]	0.04	2.62	1.76	0.004	0.056	0.041	0.004	0.085	0.060	0.008	0.184	0.129	7.33
Dolomitic limestone[c]	0.06	0.30	0.26	0.004	0.010	0.011	0.007	0.016	0.017	0.018	0.059	0.057	21.98
Siliceous limestone[d]	0.11	0.24	0.27	0.009	0.003	0.010	0.012	0.002	0.013	0.011	0.003	0.013	5.00
Pure limestone[d]	<0.10	0.09	<0.16	0.003	0.003	0.005	0.013	0.010	0.019	0.007	0.015	0.017	14.41

(continued)

Table 3.5 (continued)

	Total alkali content (wt%)			Soluble alkali content from the aggregate (wt%)									
				Distilled water			Lime oversaturated			Alkaline solution (0.7N)			MAX
Aggregate lithology	Na_2O	K_2O	Na_2O_{eq}	Na_2O	K_2O	Na_2O_{eq}	Na_2O	K_2O	Na_2O_{eq}	Na_2O	K_2O	Na_2O_{eq}	wt%/ total
Siliceous limestone[e]	0.04	0.08	0.09	0.003	0.003	0.005	0.003	0.003	0.004	0.006	0.005	0.009	9.24
Silicified sandstone[f]	0.01	<0.05	<0.04	0.001	0.008	0.007	0.001	0.006	0.005	>0.002	>0.014	>0.011	>25.60
Quartzite	<0.01	<0.05	<0.04	0.000	0.001	0.001	0.001	0.001	0.002	>0.002	>0.002	>0.004	>8.05
Calculated average	2.35	1.75	3.50	0.026	0.016	0.036	0.028	0.022	0.042	0.079	0.058	0.117	>7.04

[a] Lithic gravel is from Sudbury, [b] Dolostone is from Beekmantown, [c] Dolomitic limestone is form Kingston, [d] Siliceous limestone and Pure limestone are from Trenton, [e] Siliceous limestone is from Spratt, [f] Silicified sandstone is from Potsdam

These authors found an alkali contribution to concrete, after 578 days, varying from 0.1 to 3.4 kg·m^{-3} Na_2O_{eq}, excluding the much higher value of 12.7 kg·m^{-3} for a sample of phonolite. In general, much more alkalis were released in alkaline solutions than in lime saturated solution, with lower values obtained in water. In the same work, the authors found that limestone aggregates ground to 0.63–1.25 mm release about 15–20% more alkalis, in a 3 months test, than with the size of 1.25–5 mm and suggest that alkalis are easy to remove from the interior of particles <5 mm. Way and Cole [260] showed that at 50 ºC some aggregates released larger amounts of alkalis in CH saturated solutions when finely crushed than in coarsely ground fractions. Another important conclusion concerns the type of aggregate: aggregates of sedimentary origin released much less alkalis than of metamorphic or igneous origin.

- Ouali [185]
  In this study, 7 of the 8 aggregates selected by Le Roux et al. [153] were leached in 500 mL of 0.7N NaOH and KOH solutions stored at 60 °C. A sub-sample of 120 g with a particle size ranging from 0.08 to 0.63 mm was used. The alkaline solution was chosen since it mimics the concrete pore solution which contains a low calcium content (crystallized in portlandite) and because the attack with lime water can lead to the precipitation of secondary minerals that can disturb the alkali released measurement. However, the experimental conditions originally from Le Roux et al. [153] (LCPC method number 37 [151]) (solid/liquid ratio 1: 4.2, T = 60 °C, 1000 h duration) were changed to 1:1, T = 100 °C, and 7 h duration, respectively.
  The soluble alkali contents measured ranged from 0.008 to 0.47 wt% Na_2O_{eq} (mean 0.094 wt%). The results of Ouali [185] showed that the release of alkali by the aggregates decreases by 5 to 10 times for experiment run at 20 °C compared to 60 °C. In addition, alkali release was higher in the short term and lower in the long term at a low solid/liquid ratio. Finally, more alkalis were released when the aggregates were immersed in alkaline solutions rather than in a lime saturated solution. These results are in agreement with those of Bérubé et al. [10].

- Wang et al. [257]
  Wang et al. [257] analysed non-reactive alkali-bearing aggregates mixed with reactive aggregates. They tested nepheline syenite and alaskite (=light coloured *alkali-feldspar-granite*) using an accelerated test method where aggregates with particle size ranging from 0.15 to 0.65 mm were immersed in alkaline and in lime saturated solutions for 24 h at 150 °C. Authors concluded that Na-feldspar and K-feldspar are stable in alkaline solution with slight dissolution at the crystals interface. The release of alkalis results from the ion exchange with Ca^{2+} in solution. It was also found that these aggregates increase ASR expansion in a more evident way when low alkali cement is used and that the maximum released alkali content was increased 40× when the pore solution/aggregate ratio increased from

0.05 to 2. The alkalis release from the studied minerals was considered directly dependent on the saturation condition of the solution.

Nepheline-bearing aggregates proved to enhance ASR in concrete [89, 90]. This conclusion was also found by Durand [66] when using non-reactive nepheline syenite with some reactive aggregates in expansion tests, in result of synergetic effect between the aggregates.

Regarding the pore solution, Wang et al. [257] found that the contribution of alkalis from the aggregates decreases when high alkali cements are used, possibly due to saturation of the pore solution. Also, Wang et al. [258] concluded that the use of low water/cement ratio and the prevention of infiltration of water into the concrete would decrease the release of alkalis from the aggregates due to an increment in the alkali concentration in the solution, therefore decreasing the reaction with the aggregates to release alkalis.

- Locati et al. [157]
  Locati et al. [157] studied two sodium feldspars and two potassium feldspars with particle size between 0.15 and 0.30 mm, immersed in 1 M alkaline solutions at a temperature of 80 °C. The soluble alkalis measured ranged from 0.009 to 0.016 wt% Na_2O_{eq}.

- Berra et al. [7]
  Berra et al. [7] investigated the alkali release potential of a potentially reactive carbonate sedimentary rock and an unreactive extrusive altered feldspathic rock. The test was carried out on two particle size fractions, namely from 0 to 4 mm and from 4 to 22.5 mm, as used during concrete manufacturing. The aggregate particles were immersed at 80 °C for 180 days in alkaline solutions with a concentration of 0.7M NaOH and 0.7M KOH at a solid to liquid ratio of 1:1. The fine fraction released 0.037 wt% and 0.041 wt% soluble Na_2O_{eq} for the sedimentary and extrusive rocks, respectively. For the coarse fraction, the soluble alkalis were 0.011 wt% and 0.015 wt% Na_2O_{eq}, respectively.

- Soares et al. [219]
  In this study, the alkali release of five granitic aggregates commonly used for the manufacture of concrete was evaluated. Aggregate particles were immersed in saturated lime solution, in 1M KOH (Na measurement), and 1M NaOH (K measurement) with solid lime in excess. Three size fractions were studied, namely 0–2 mm, 4.75–9.5 mm and 12.5–20 mm, at an aggregate to liquid ratio of 1:4. The tests were conducted at 38 °C. After 7, 14, 28, 56, 91, 180, 365 days, 25 mL of solution were taken for chemical analysis. The same volume of stock solution was subsequently reintroduced and returned back to the test conditions. In the saturated lime solution at 365 days, the release of alkali by the aggregates ranged from 0.024 to 0.045 wt% Na_2O_{eq} for the particle size fraction 0–2 mm, from 0.007 to 0.010 wt% Na_2O_{eq} for the fraction 4.75–9.5 mm, and from 0.005 to 0.007 wt% Na_2O_{eq} for the fraction 12.5–20 mm. In alkaline solutions at 365 days, the release

of alkali by aggregates ranged from 0.190 to 0.206 wt% Na_2O_{eq} for the 0–2 mm fraction, from 0.131 to 0.153 wt% Na_2O_{eq} for the 4.75–9.5 mm fraction and from 0.104 to 0.144 wt% Na_2O_{eq} for the size fraction 12.5–20 mm.

Results showed that the finer particle size fraction released more alkali than the coarser fraction. Also, to compare these results with previous studies, the same experiment was carried out on one of the five aggregates using 0.7M KOH and 0.7M NaOH as immersion solutions. The latter showed that the 0.7M KOH solution was more effective at extracting Na^+ than 1M concentration.

- Menéndez et al. [166]
  In the study, the potential extraction of alkalis from a granodiorite is analysed by six different methods of extraction in alkali solutions. Two of these methods are based on saturated calcium hydroxide. These solutions are used in the conditions showed in the Table 3.6.

  The results of extraction, expressed as Na_2O_{eq}, are 0.15 wt% and 0.12 wt%, respectively, and the most efficient extraction solutions are these, with respect to the solutions with sodium, potassium or sodium plus calcium and potassium with calcium. In this case, the role of the calcium would be related to the ionic exchange with sodium and potassium, which could be brought about in the concrete, given the elevated content of calcium in its liquid phase.

  Menéndez et al. [166] have proposed to use a solution that simulates the pore solution of concrete, with sodium, potassium and calcium (0.1N NaOH + 0.6N KOH + $Ca(OH)_2$ sat.) [165]. This solution is more similar to the situation of the aggregates in the concrete. In that study, the potential extraction of alkalis from a granodiorite is analysed at 20 °C and 40 °C, during 28 days and liquid: solid ratio of 3:1. The results show that the extraction appears similar to the behaviour of the granodiorite in a field concrete, after fifty years of exposure to a high water environment. The usual extraction procedure using two solutions (NaOH, KOH) to extract two alkalis (K, Na) tends to overestimate the contribution from this granodiorite. However, interpretation of analytical results from the more complex solution is challenging due to interactions between the dissolved species from different reservoirs (see e.g., [22]).

Table 3.6 Methods used for the potential extraction of alkalis with $Ca(OH)_2$ saturated solution

Method	Size of the aggregate	Liquid: solid ratio	Temperature	Time	Extraction solutions
Method 1 [165]	315 μm <aggregate> 160 μm	3:1	20 °C and 40 °C	28 days	D3 → $Ca(OH)_2$ Sat
Method 2 [257]	150 μm <aggregate> 650 μm	1:1	150 °C	24 h	R3 → 0.7N KOH

3.2.7.3 Alkali Extraction by Complexing Solutions

- St. John and Goguel [225]
  0.25 g of basalt powder (<0.6 mm) was immersed in 35 mL of 0.2M NH_4EDTA and 1M HNO_3 for a period of one hour. The alkali released content measured were 0.015 wt% and 2.482 wt% Na_2O_{eq} for the EDTA and HNO_3 solutions, respectively.

- Goguel and Milestone [95]
  Goguel and Milestone [95] found that by increasing the w/c ratio of the concrete mixture the percentage of the alkali taken up by hydrates decreases significantly. The authors tested nepheline basanite and concluded that large concentrations of alkalis were released when curing mortars under high humidity leading to high pH even when low-alkali cement was used.

3.2.7.4 Pore Pressing

- Constantiner and Diamond [49]
  Constantiner and Diamond [49] made mortar specimens using an ASR reactive flint as main aggregate with 25 wt% replacement of finely ground microcline, an alkali-bearing feldspar. For comparison, three different series of similar mortars were made: one with an alkali-free, almost pure limestone with the same 25 wt% replacement of finely ground microcline, another with the same ASR reactive flint, but without replacement; and finally the pure alkali-free limestone without replacement (control). The mortar pore solutions were extracted using the high-pressure extraction method. Results showed that the alkali concentration in the pore solution was constant over time for specimens without ASR reactive or alkali-bearing aggregates. In the presence of ASR, the alkali concentration was significantly lower than the control and decreased over time. The alkali concentration in the specimens containing the limestone with microcline was increasing constantly over time. For specimens containing both reactive and alkali-bearing aggregates, the alkalis released by microcline outweighed the removal of alkalis by the formation of ASR products.

  Constantiner and Diamond [49] concluded that the released alkalis could be removed from the solution by reactive aggregates therefore participating in ASR and forming reaction products. The authors also found that between one-third to one-half of the alkalis released by feldspar were removed from the pore solution and incorporated into ASR products.

- Drolet et al. [61]

  Drolet et al. [61] presented a study on the validation of the alkali contribution by aggregates to the concrete pore solution. These authors studied concrete and mortar specimens made with reactive and non-reactive aggregates, with or without alkalis. The alkali content of specimens was determined over time using the pore pressing and expresso hot water extraction methods. One aggregate was clearly identified as a potential source of Na^+ to the pore solution. Results obtained after one year at 60 °C showed that rich alkali-bearing aggregates can release up to 1.75 kg Na_2O/m^3 of concrete if used as fine fraction (0 to 4 mm) and up to 0.24 kg Na_2O/m^3 of concrete if used as coarse fraction. However, no clear K^+ contribution from aggregate was observed.

3.2.7.5 Summary of Alkali Extraction Methods for Aggregates and Influencing Factors

The release of alkalis in field concrete was confirmed by Bérubé et al. [11] who measured the water-soluble alkali content at different depths inside the concrete of several hydraulic structures. The authors concluded that the values exceeded the anticipated soluble-alkali content from the cement after hydration, and therefore the aggregates were supplying alkalis to the concrete pore fluid, thus explaining why structures built with low-alkali cements still were affected by ASR [8]. The paper from 2000 also confirmed earlier research by Nixon et al. [181] and Stark [223] who observed alkali enrichment towards the surface where the pore solution evaporates (sometimes referred to as 'chimney effect' [111]), and therefore a proportionally lower apparent soluble-alkali content. Alkalis can precipitate near surface through evaporation, as observed in the wall of a gallery of a large concrete dam (Fig. 3.8), or eventually be leached out by rain or running water. Also, alkalis from external sources (deicers, marine water) may leach in via cracks to the very center of bridge decks of Broekmans [22, 23].

In short, these field observations show that many factors influence the release of alkali by aggregates. A summary of all the experiments discussed above is presented in Table 3.7, including the parameters used in the tests and the main results obtained.

In general, the higher the pH of the immersion solution, the more aggregates will release alkalis. Also, the release of alkalis is greater when the specific surface of the granular material under test is high (fine particle size). Temperature plays an

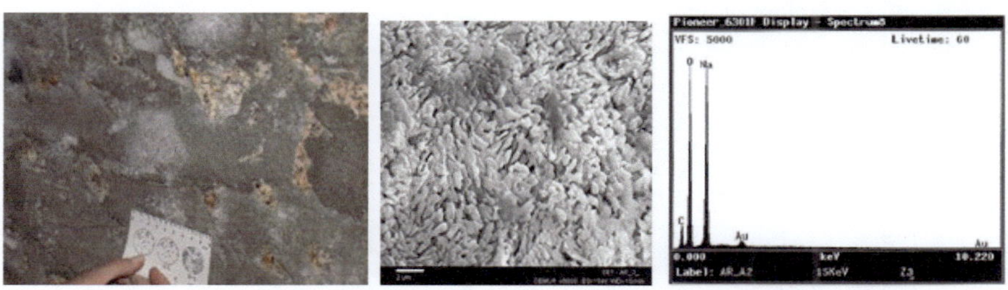

Fig. 3.8 Wall of a gallery in a concrete dam with efflorescence, SEM image of the efflorescence in SE (Secondary Electron) mode, and corresponding EDS spectrum [73]

Table 3.7 Laboratory assessment of alkali release by aggregates (updated from Bérubé and Fournier [15], adapted from Drolet et al. [61])

Sources	Material	Particle size	Test Mass (g)	Duration (d, h)	T (°C)	Agitation	Extraction solution	Ratio to solid	Released alkalis (wt% Na_2O_{eq})
Van Aardt and Visser [245]	6 feldspars, 2 clay minerals	'Powder'	0.5	250d	39	N	25 mL saturated lime[a]	1:50	0.02–2.32
Stark and Bhatty [224]	3 feldspars, 3 sands, 3 gravels, 1 andesite	<80 μm	5	90d	38	Y	25 mL saturated lime[a]	1:5	0.26–1.99
	3 feldspars, 1 andesite	<80 μm	5	28d	80	N	25 mL saturated lime[a]	1:5	0.83–5.03
							25 mL distilled water		0.03–0.13
Kawamura et al. [132]	2 andesites	0.15–5 mm	100	180d	40	Y	25 mL saturated lime[a]	1:2	0.02–0.07
Bérubé et al. [10]	17 aggregates	1.25–5 mm	40	578d	38	Y	40 mL distilled water	1:1	0.001–0.151
							40 mL saturated lime[a]		0.002–0.11
							40 mL alkaline solution[b]		>0.004–0.684
Le Roux et al. [153]	8 aggregates	0.08–20 mm	500	7h	100	N	500 mL saturated lime[a]	1:1	0.0085–0.21

(continued)

Table 3.7 (continued)

Sources	Material	Particle size	Test Mass (g)	Duration (d, h)	T (°C)	Agitation	Extraction solution	Ratio to solid	Released alkalis (wt% Na_2O_{eq})
Ouali [185]	7 aggregates	0.08–0.63 mm	120	1000h	60	N	500 mL alkaline solution[b]	1:4.2	0.008–0.47
Lu et al. [163]	Gneiss, Granite, K-feldspar	10–12 mm	25	28d	80	Y	50 mL saturated lime[a]	1:2	0.006–0.016
		5–10 mm							0.007–0.029
		1.25–5 mm							0.011–0.037
		0.63–1.25 mm							0.015–0.046
		0.15–0.63 mm							0.022–0.064
		0.08–0.15 mm							0.051–0.112
Wang et al. [257]	Nepheline, Alaskite	0.15–0.65 mm	unknown	24h	150	N	saturated lime	1:1	–
							alkaline solution[b]		–
Locati et al. [157]	2 K-feldspars, 2 Na-feldspars	0.15–0.30 mm	25	24h	80	N	alkaline solution[b]	1:10?	0.009–0.016
Bérubé et al. [7]	2 aggregates	0–4 mm	unknown	180d	80	N	alkaline solution[b]	1:1	0.037–0.041
		4–20 mm					alkaline solution[b]		0.011–0.015

(continued)

Table 3.7 (continued)

Sources	Material	Particle size	Test Mass (g)	Duration (d, h)	T (°C)	Agitation	Extraction solution	Ratio to solid	Released alkalis (wt% Na_2O_{eq})
Soares et al. [219]	6 granitic aggregates	0–2 mm	unknown	455d	38	Y	saturated lime	1:4	0.024–0.045
		4.75–9.5 mm					saturated lime		0.007–0.010
		12.5–20 mm					saturated lime		0.005–0.007
		0–2 mm					alkaline solution[c]		0.190–0.206
		4.75–9.5 mm					alkaline solution[c]		0.131–0.153
		12.5–20 mm					alkaline solution[c]		0.104–0.144
Menéndez et al. [166]	Granodiorite	0–4 mm	Unknown	91d	60	unknown	alkaline solution[b]	1:4	–
Drolet et al. [61]	2 limestones, 2 greywackes, 1 phonolite, 1 granitic gneiss	0–4 mm	200	365d	38 and 60	Y 2× week 800 mL	alkaline solution[b]	1:4	0.02–0.30
									0.33–0.47 (high alk agg)

[a] Saturated lime solution with solid lime in excess
[b] 0.7M KOH to measure Na release and 0.7M NaOH to measure K release
[c] 1M KOH to measure Na release and 1M NaOH to measure K release, lime saturated

important role in the kinetics of the reaction with faster and higher alkali release with the increase of the temperature [15]. The presence of cleavage planes or microcracks, which can be attributed to retrograde metamorphism and/or fracturing of the concrete, favours access to the attack solution, thus allowing better alkali release [26]. Finally, the weathering condition or alteration of the aggregates also influences the release of the alkalis as shown by Ferraz et al. [77]. This subject is presented in the next section.

The test procedure that is most adequate for the assessment of the amount of alkalis released by aggregates is yet to be determined, mainly in what concerns the sample preparation, the solution used for the extraction of alkalis, the duration and temperature of the test [163].

Menéndez et al. [167] have analysed the different parameters used in the tests to assess the potential alkali release form aggregates and their contribution to ASR. Comparing the different methods of extraction, it is observed that the main parameters are temperature, duration, type and concentration of extraction solutions, liquid/solid ratio, particle size and granulometry, sample size, agitation of sample and reposition of solution (Fig. 3.9).

Alkali-release tests are in principle designed to reflect the conditions present in field structures, though a certain degree of acceleration is necessary to avoid excessively long test durations that are impractical.

According to experiments by many researchers over several decades, it appears most efficient to evaluate the releasable alkali content by using saturated lime solutions ($Ca(OH)_2$) at 80 °C, with a 1:5 or 1:2 solid: liquid ratio, for 28 days. Certain aggregates have been found to release up to 5 wt% Na_2O_{eq}. How and to what extent these test results represent field concrete is still a matter of discussion.

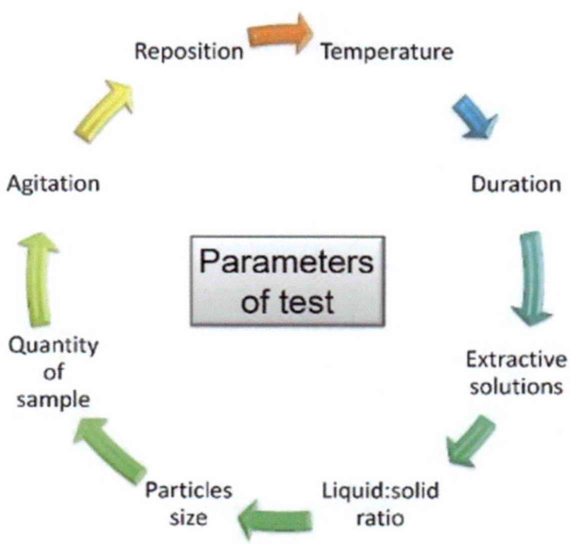

Fig. 3.9 Parameters that control the extraction of alkalis [167]

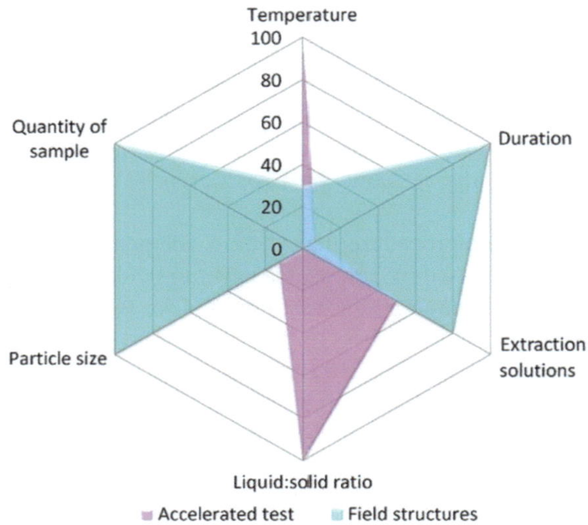

Fig. 3.10 Relative values of the different parameters between accelerated tests and field concrete [167]

Methods that apply non-accelerated test conditions are arguably more realistic, and the amounts of released alkalis are perhaps more representative for field concrete, but require very long test duration (Fig. 3.10).

Long-term assessment of the behaviour of aggregate materials in large critical structures like hydropower dams and nuclear power plants requires numerical modelling of released alkali values and their contribution to initiate deleterious alkali-silica reaction [167].

3.2.7.6 The Effect of Alteration of Rock Lithologies

Aggregate samples for laboratory testing need to be representative of the material as used in concrete (See Sect. 3.1). To compensate for natural intra-deposit variation (e.g., local differences in weathering, faulting/fracturing, sedimentary/metamorphic facies; [74]), quarries tend to mix materials from different extraction faces to obtain a homogenous final product with only little variation.

Recently, Ferraz et al. [77] assessed the effects of weathering on alkali-release in seven samples of granitic aggregate, by handpicking fresh and visually unaltered particles from altered particles into separate subsamples (also see [207]). The degree of alteration was assessed by optical thin section petrography, using image analysis on photomicrographs to identify alteration of K-feldspars to sericite and kaolinite and of plagioclase to saussurite, and to reveal the presence of cracks fragmenting K-feldspar and quartz crystals (Fig. 3.11). Granite with stronger alteration generally showed a higher degree of deformation than unaltered granite from the same deposit.

To determine the amount of releasable alkalis, all materials were comminuted to pass a 150 μm sieve. Then, 100 g of powder was immersed in 400 mL of alkaline

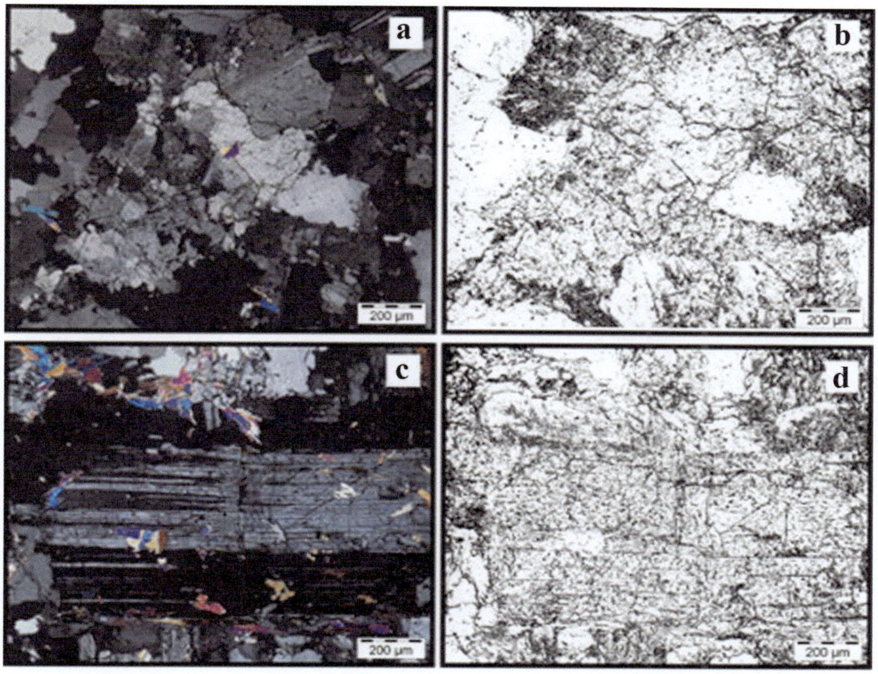

Fig. 3.11 Optical micrographs reveal mineral alteration and cracking [77] **a**, **c** in cross-polarized light; **b**, **d** in plane-polarized light

solution (0.7M NaOH to assess K release; 0.7M KOH to assess Na) at a ratio of 1:4 solid/solution and kept at 38 °C. Test vessels were agitated by gentle rolling back and forth for about 10 s twice a week, and 10 mL of solution extracted as prescribed for assay of Na and K by atomic absorption spectroscopy (AAS) at predefined time intervals 2, 7, 13 and 26 weeks.

Detailed assessment of the exposed materials revealed that plagioclase is always more altered and cracked than K-feldspar, both in the visually unaltered (fresh) as well as in the altered subsamples. The amount of potassium released was larger than the amount of sodium, which is attributed to the greater abundance of K-feldspar in granite compared to plagioclase, as well as to the higher initial degree of natural alteration of the plagioclase. In conclusion, natural alteration of aggregate (constituents) reduces their capacity to release alkalis compared to their unaltered equivalents, as shown in the left panel in Fig. 3.12 (see next page). This confirms the findings of Way and Cole [260] from 1982, who state that aggregate of sedimentary origin (from which alkali-rich minerals like K-feldspar, plagioclase, mica, etc., were removed by weathering, erosion, sedimentary transport and deposition) released much less alkali than aggregate from metamorphic or igneous rock types.

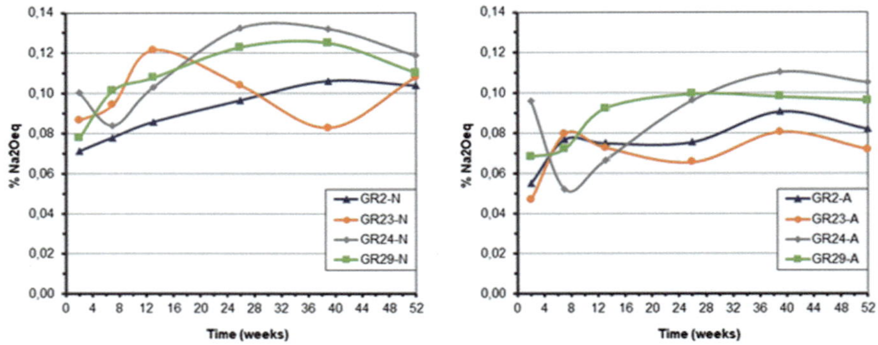

Fig. 3.12 Alkali release of four samples, comparing values for unaltered (N; left panel) and altered (A; right panel) subsamples [207]

3.2.8 Extraction of Alkalis from Supplementary Cementitious Materials

All supplementary cementitious materials - SCMs contain Na and K, some of them even more than the OPC they replace (most notably fly ash). This has raised questions about the alkalis availability for deleterious ASR, and whether or not they should be included in the calculated total-alkali content of a given concrete mix. In hardened concrete, alkalis occur dissolved in the pore solution, consolidated in clinker and its hydration products, and bound in alkali-reaction products, each with different degrees of mobility. In SCMs, the alkalis are bonded by the glassy particles, and are released upon their hydration in the paste. The role of alkalis in fly ash and their contribution to the pore solution in concrete has been studied by e.g., [116, 180, 237]. The key reason to add SCMs to a cement is that they reduce the potential for ASR damage by immobilizing alkalis, thus reducing their availability in the concrete pore solution. If all aggregate is innocuous (i.e., *not* alkali-reactive), then the partitioning of alkalis between pore solution and clinker hydration products is a simple function of the binder (=OPC + SCMs) composition [239].

Shehata and Thomas [213] showed that the efficacy of fly ash to prevent expansion is directly related to its composition, notably CaO and Na_2O_{eq} contents. Fly ash with >5 wt% Na_2O_e is ineffective at 25 wt% OPC substitution, whereas Class F fly ash with Na_2O_e at 8.45 wt% and CaO at 12.3 wt% is ineffective even at 40 wt% substitution. Carrasquillo and co-workers [32, 33] observed a pessimum effect, which they attributed to alkali release from fly ash with >1.5 wt% Na_2O_{eq} at high substitution levels. A review of different types of SCMs analysed by fourteen different methods [168] confirms that release of alkalis over time is very low, the largest amounts occurring in pozzolanas and limestones, the lowest in silica fume. According to [44], SCMs remain effective in suppressing ASR expansion even when alkalis are supplied by external sources, hereunder release from aggregates.

The alkali contents of SCMs are regarded differently from country to country in national guidelines:

- The Concrete Society [46] and Building Research Establishment in the UK [20] do not require the alkali content of blastfurnace slag or fly ash to be included in the calculation of concrete alkali content. For blastfurnace slag, none of the alkali is included if it comprises more than 41 wt% of the binder. 50% of its alkali-content is included if the blastfurnace slag comprises 25–41 wt% of the binder, and all alkali is included if the blastfurnace slag comprises less than 25 wt% of the binder.

  For fly ash, none of the alkali is included in the calculation if the fly ash comprises more than 25 wt% of the binder. Only 20 wt% is included if the fly ash comprises 20–25 wt% of the binder, and all the fly ash alkali is included if the fly ash comprises less than 20 wt% of the binder. These require that the acid-soluble or total alkali content of blastfurnace slag does not exceed 1.0 wt% Na_2O_{eq} and fly ash does not exceed 5.0 wt% Na_2O_{eq}. BRE [20] places no limit on the alkali content of silica fume. The alkali content of metakaolin is disregarded.
- The concrete alkali limits in Canadian Standard CSA A23.2–00 [51] only take into account alkali contribution from the Portland cement fraction, but the total alkali content of SCM is taken into account when determining cement replacement levels. SCM with higher total alkali content can be used if testing shows it to be effective in reducing ASR expansion.
- The Cement and Concrete Association of Australia [39] considers that 50 wt% of the total alkali in blastfurnace slag, 15–20 wt% of that in fly ash and most alkalis in silica fume contribute to the alkali content of pore solution and must be taken into account. It is recommended that the alkai content of fly ash should be limited to a maximum of 3 wt% and total alkali content of blastfurnace slag and silica fume should be limited to a maximum of 1 wt%, with approximate levels of replacement of 10 wt% for silica fume, 25 wt% for fly ash and 65 wt% for blastfurnace slag.
- The ASTM specification is based on the effectiveness of reaction mitigation using SCMs on the expansion mortar accelerated bar tests, for blastfurnace slag, fly ash, natural pozzolan and silica fume [40].
- The French standard NF P18-454 [178] considers different percentages of alkalis for different SCMs, in order to calculate the content active alkalis in the cement. In this standard the percentages of effective alkalis in different SCMs or other components of the binder are as follows: 100 wt% for clinker (K), 100 wt% for gypsum (G), 50 wt% for slag (S), 17 wt% for the silica fume (D), 17 wt% for the natural pozzolanas (P), 17 wt% for the calcined natural pozzolanas (Q), 17 wt% for the siliceous fly ash (V), 100 wt% for the calcareous fly ash (W), 100 wt% for calcined schists (T), 50 wt% for limestone (L) and 50 wt% for calcined limestone (LL). These percentages shall be applied as a weighting factor for the percentages of substitution of SCMs and based on the content of alkalis in each SCMs.

According to [32, 33] a high substitution of fly ash with high alkali content (>1.5 wt%) produces a pessimum effect in ASR, which is associated with the release of alkalis from the fly ash.

Some of the alkalis in fly ash are consolidated in the glassy particles and are released when the fly ash reacts in concrete to form calcium silicate hydrates. A number of researchers have studied the role of fly ash alkalis and their net contribution to the alkalinity of the pore solution in concrete corroborating this effect [116, 180, 237].

However, experiments with different types of SCMs analysed by fourteen test methods reveal that the amounts of alkalis released with time are very low [168]). The maximum alkali release was observed in limestones and pozzolans, the lowest in silica fume.

3.3 Instrumental Methods to Determine Alkali Contents and Other Constituents in Bulk Materials and In-situ

Traditional methods of chemical analysis are numerous and based on a very diverse range of techniques that use a wide variety of physical phenomena. The term "chemical analysis" is used to refer to any measurements that provide information about the chemical composition of a sample [88, 235]. A wide range of elemental analysis techniques are used to identify and quantify the elemental composition of materials and compounds. Qualitative chemical analysis refers to the techniques used to determine the identity of the elements present in an unknown sample while the quantitative chemical analyses allow the determination of the amount by weight of each element or compound present.

Element species present in rocks, minerals or related solids are divided into three different categories depending on their abundances in a sample. In the analysis of "earth materials", major (main) elements have concentrations >1.0 wt%, and their analytical sum total comprises > 95 wt% of the sample. Which set thus composes 'the major elements' varies with the nature and bulk composition of the analysed material, so that a limestone has a different set of major elements than a granite, an OPC clinker, or an SCM. In bulk whole-rock, minerals and related geomaterials (e.g., aggregate, clinker) analysis, major element contents are usually stated as weight percentages (wt%) of oxides, e.g., Na_2O, K_2O, MgO, CaO, meaning that the content of each element analysed is recalculated into its stoichiometric oxide. Iron content can be listed as either FeO^T or $Fe_2O_3{}^T$ to represent total-iron content, unless ferrous and ferric iron each have been measured separately (which then would be specified). In certain minerals (-groups), ferrous and ferric iron contents may be calculated based on a number of assumptions [63].

For certain types of analyses, volatile content is given as Loss on Ignition - LOI, also in wt%. The (unnormalized!) analytical sum total of a satisfactory bulk whole-rock analysis including LOI should be within 100.0 ± 1.50 wt%-sum totals outside of this range are a strong indication that an important constituent was omitted. This is different for SEM–EDS and/or EPMA analyses, which is elaborated in further detail in Sect. 3.4.

Minor elements have concentrations >0.10 wt% or >1 000 parts per million (ppm; $\mu g \cdot g^{-1}$, or $mg \cdot kg^{-1}$, or $g \cdot tonne^{-1}$), and trace elements have concentrations below 1 000 ppm - there is no bottom limit so (ultra-) trace contents may be reported at ppb (parts per billion, $ng \cdot g^{-1}$), ppt (parts per trillion, $pg \cdot g^{-1}$), ppq (parts per quintillion, $fg \cdot g^{-1}$), even down to single atoms being detected. Minor and trace elements are normally not recalculated to oxides.

The distinction major-minor-trace is rather arbitrary and their transitions gradual, as opposed to strict. Element species may change roles from result to result in a given sample set and be reported as wt% among the major element oxides, as well as in ppm when listed together with minor and trace elements. How analytical results are reported is of major influence on the correct understanding of the data, as elaborated by [85] for the example of manganese Mn.

The concept of major-minor-trace elements is redundant for hydrous solutions, as *the* main 'major constituent' by a very large extent would be the solvent, water. Note that contents of hydrous preparations for instrumental analysis (e.g., a sample digested in excess strong acid) are calculated to represent element (oxide) contents as present in the initial solid.

An important part of all analytical procedures is the calibration and standardization process [216]. Most geochemical methods require calibration using certified reference materials. Reference material is defined as a material sufficiently homogeneous and stable with respect to one or more specified properties to be used for the calibration of an apparatus or the assessment of a measurement method. These are quality control samples that are used to document changes in measurement results within and between batches. The reference materials that are characterized in a given laboratory are often called internal standards whereas those characterized by an independent organization are called certified reference materials. These standards are essential for assessing the accuracy and precision of an analytical method. Reference materials are chosen to closely match the matrix and analyte concentration of interest. For this purpose, one of the best existing compilations of certified reference materials is that in Jochum et al. [130].

Analytical methods are often distinguished in classical or instrumental methods. This classification is largely historical with the classical methods sometimes called wet chemical methods, preceding 'instrumental methods'. The main methods of chemical analysis are described below. The purpose of the text is to present the basic concepts in analytical geochemistry and is not intended to be an exhaustive text on the subject. Several excellent works deal with this subject in detail. The aim is to present the basic principles of important publications [38, 79, 88, 114, 129, 156, 176, 184, 192, 193, 208, 216, 262].

The approach of chemical analysis is to measure a signal related to the concentration of an elemental constituent (analyte) in a sample. This analyte signal invariably has a background noise component, the magnitude of which usually limits the lowest concentration to be measured satisfactorily (the limit of detection). These concepts are developed for the main analytical chemistry techniques presented below.

3.3.1 Wet Chemical Analysis

In the early years of chemistry, most measurements were made by separating the element of interest in a sample by precipitation, extraction, or distillation. For qualitative analyses, the separated components were then treated with reagents that formed products recognizable by colour, boiling or melting point, solubility in a series of solvents, odour, optical property, or refractive index. For quantitative analyses, the measurements are based on chemical stoichiometry measured with gravimetric methods (measurement of the mass of a certain compound produced from the element) or titrimetric measurements (measurement of the volume or mass of standard reagent required in order to react completely with the element) [216]. Another technique is colorimetry that is used to determine the concentration of coloured solutions by the application of the Beer-Lambert law, which states that the concentration of a solute is proportional to the absorbance [187]. The colorimeter is a device that measures the absorbance/transmittance of particular wavelengths of light by a sample in solution that is compared with the absorbance of standard solutions.

Advantages and Limitations

Provided equipment and instrumentation are already available in-house, wet-chemical analysis of selected element species may appear cheaper compared to ordering instrumental analysis from an external supplier, and turn-around time will generally be shorter.

However, including cost for laboratory facilities, trained personnel, annual service and maintenance etc., quickly tips the balance in favor of instrumental analysis, especially when these provide data for up to forty elements per sample, decreasing the cost per species.

3.3.2 X-ray Fluorescence (XRF)

XRF is a relatively non-destructive sensitive analytical technique used to determine the elemental composition of inorganic materials. The XRF method is described in detail by several authors including Jenkins [129], Nakayama and Nakamura [176], Van Grieken and Markowicz [249], Margui and Van Grieken [164] and others.

XRF is used for bulk chemical analysis on different kind of materials including liquids, solids, loose powders, pressed pellets, fusion beads, pastes and thin films samples (See e.g., [29]). XRF is based on the principle that atoms constituted in the sample pass from their ground state to an excited state when placed under a high-energy X-ray beam. The excited state is unstable and the atoms tend to return to the ground state by releasing energy in the form of secondary (fluorescent) X-ray photons. Each atom, having a proper electronic configuration, will emit fluorescent X-ray photons with a characteristic energy and wavelength. The fluorescent radiation can be analysed either by sorting the energies or by separating the wavelengths of the photons. The intensity of each characteristic radiation is directly related to the amount of each element in the material. Two categories of XRF instruments are defined depending on the type of spectrometer used for the detection of the X-ray photons emitted: energy-dispersive X-ray fluorescence (EDXRF), and wavelength-dispersive X-ray fluorescence (WDXRF). Each has a different detection system, different instrument characteristics, and different application areas.

EDXRF is a convenient technology to screen all kinds of materials for quick identification and quantification of elements with little or no sample preparation. Instruments with EDS are generally equipped a with solid-state detector and multi-channel pulse-height analyser, the obtained XRF spectrum is usually displayed with X-ray energy in keV along the abscissa [176]. EDS instrument types have the great advantage to display information on all elements at the same time, but lack in resolution over the WDS [129]. EDS can be used for monitoring major components in an industrial product or process or the addition of an additive.

Wavelength-dispersive spectrometer (WDS) is based on the principle of Bragg diffraction of the fluorescent X-rays emitted by the material sample that is diffracted on a single crystal before being detected. A detector is angularly scanned (with a rotating goniometer) relative to the analysing crystal according to the Bragg's Law, registering the spectrum. The WDXRF has superior resolution and sensitivity compared to EDXRF. The WDS is superior to EDS for the determination of light elements as Na and Mg. WDS type instruments are commonly used to produce excellent result for whole rock analysis of the fourteen major oxides (Na_2O, MgO, Al_2O_3, SiO_2, P_2O_5, K_2O, CaO, TiO_2, MnO, SO_3, Cr_2O_3, ZnO, SrO, and total Fe_2O_3) [176].

According to Jenkins [129], EDS and WDS differ mainly in the excitation source, the number of elements measured at one time, the speed of data collection and their price. Theoretically, both EDXRF and WDXRF instruments have the ability to detect X-ray emissions from all elements in the periodic table from $Z = 9$ (F) and upwards, and WD spectrometers can measure down to $Z = 4$ (Be). Benchtop and mainframe instrument types can be fitted with automated sample-feed [129]. However, in practice, most commercially available instruments are limited in their ability to precisely and accurately measure the abundances of elements with $Z < 11$ (Na) in most natural earth materials. According to Nakayama and Nakamura [176], XRF spectroscopy may reach levels of 10–100 μg/g (ppm) to 100 wt% for elements from Be ($Z = 4$) to U ($Z = 92$). Older XRF systems and/or recent budget-type or portable instruments are restricted to analysing only Na (11) to U (92). XRF is generally unsuitable for ultratrace analysis below ppm (μg/g) concentration levels.

While instrumental precision in EDXRF systems has increased greatly in the last decade, WDXRF instruments are much more precise particularly in the light elements such as Na or lighter [208]. According to Jenkins [129], the detection limits for Na_2O measurements are 0.81 and 0.16 wt% for EDXRF and WDXRF, respectively. For K_2O, values of 0.022 (EDXRF) and 0.008 wt% (WDXRF) were reported. Because of its lower resolution at low energies, it is difficult to determine precise light element abundances (particularly Be (4) to Cl (17)), with an EDS system in comparison to a WDS system [176].

According to Jenkins [129], the absolute sensitivity of XRF decreases dramatically with low atomic number elements mainly due to reduction in the fluorescence yield, the decrease in the absolute number of useful long wavelength X-ray photons, and the increase of absorption effect with decreasing atomic number. The two first issues are inherent to the X-ray excitation process while the absorption problem depends on the instrument design and particularly on the detector. A combination of these factors causes a loss in the sensitivity of most EDS systems of almost an order of magnitude, for the K lines of sulfur (Z = 16) to sodium (Z = 11). In earlier years, the lightest detectable element with a typical EDS was Z = 12 (Mg). However, modern mainframe XRF instruments are capable of routine analysis of ultralight elements down to Be (Z = 4).

There is always a compromise between intensity and resolution of X-ray lines. The resolution of WDS is typically in the range of 5–20 eV, compared to 150 to 200 eV for EDS. XRF methods are sensitive with detection limits for most elements in the low ppm range up to near hundred percent. Major factors affecting the detection limit are the sensitivity of the spectrometer (counting rate), the background and the time for counting peak. The absolute sensitivity of WDS is higher than that of EDS by one to three orders of magnitude with count rates up to 1 000 000 cps (WDS) compared to 40 000 cps for EDS systems. The two methods differ mainly in their relative sensitivities with WDS roughly one to two orders of magnitude more sensitive than the EDS. However, EDS measures all elements at the same time, whereas WDS measures one line at the time on each spectrometer. Analysis time vary from 10 s to 3 min per element with minimum sample size required of a few milligrams. The accuracies obtainable are excellent mainly because the matrix effects in X-ray spectrometry are well understood and easy to overcome [129].

3.3.3 Atomic Absorption Spectrometry (AAS)

Atomic absorption spectroscopy (AAS) is an analytical technique for the quantitative determination of chemical elements mainly in solution. AAS method is described in detail by several authors including Welz [262], Potts [193], Cazes [38], Hill [114], Garcia and Baez [84] and others.

The absorption phenomenon occurs when an atom absorbs radiation whose energy corresponds exactly to a particular electronic transition required to excite an orbital electron from a stable state to an excited state [38]. The absorption depends on the presence of free non-excited atoms in the ground state [262]. Atomic absorption methods are highly specific because atomic absorption lines are remarkably narrow and because electronic transition energies are unique for each element. Because the wavelength of the monochromatic light source is characteristic only of the measured element, the light energy absorbed by the flame is a measure of the concentration of that element in the sample according to the Beer-Lambert law, the same as for colorimetry on hydrous solutions in wet chemical analysis (See Sect. 3.3.1). The concentration of the element is determined from a calibration curve using standards of known concentration [84]. The energy of light is measured by absorbance, the light energy absorbed by the sample, or by transmittance, the fraction of light that emerges from the sample [192].

In atomic absorption spectrophotometry, the sample is aspirated and atomized. The atomization process volatilizes and decomposes the sample to produce an atomic gas. The atomizer should generate as many free atoms in the ground state as possible [262]. The atomic absorption instrument is either equipped with a burner for the "flame" mode (method recognized for the low number of spectral interference) or with an electrical resistance and a graphite furnace for the "electrothermal atomizing" mode.

Flame Atomic Absorption Spectroscopy (FAAS)

Flame atomic absorption spectroscopy is currently the most widely used of all the atomic methods because of its simplicity, effectiveness, and relatively low cost. Flame AAS is widely used as a routine technique for trace element determination in the mg/L to μg/L range [262]. The sample is introduced into the excitation source in the form of a solution. The sample preparation steps need for the decomposition and dissolution are often more time consuming and introduce more error than the spectroscopic measurement itself [71, 137].

In a flame atomizer, a solution of the sample is aspirated by a capillary to be converted into an aerosol (droplets of size <5 μm) by the nebulizer. The efficiency of the nebulization is less than 10% and consequently around 90–95% of the solution ends in the drain of the injection chamber [78]. The aerosol solution is homogenized into a flame with the flow of gaseous oxidant (air or NO_2) mixed with a gaseous fuel (acetylene) where atomization occurs. A light source emitted from a hollow cathode lamp or an electrodeless discharge lamp is directed through the flame to the monochromator, and onto a detector that measures the amount of light absorbed. The absorption depends on the presence of free non-excited atoms in the ground state in the flame. Because the wavelength of the light source is characteristic only of the measured element, the light energy absorbed by the flame can be linearly correlated to the concentration of analyte present by comparison with the calibration curve. Table 3.8 and Table 3.9 presents detection limits and other characteristics for Ca, Na and K measurements in a flame atomizer.

Table 3.8 Wavelength and detection limits in a flame atomizer (from p. 20 in Van Loon and Barefoot [250])

Element	Wavelength (nm)	Detection limit (μg/mL)
Ca	422.7	0.0005
Na	589.0	0.0002
K	766.5	0.002

Table 3.9 Characteristic concentration, detection limit and normal working range in a flame AAS (from p. 169 in Hill [114])

Element	Characteristic concentration (μg/mL)	Detection limit (μg/mL)	Normal range (μg/mL)
Ca	0.01	0.001	0.01–3
Na	0.003	0.0002	0.002–1
K	0.007	0.003	0.03–2

One of the disadvantages of this method is that only one element can be quantified at a time [84]. Interferences can also affect the ground state atom population. According to Filho et al. [78], AAS is a technique with few interferences that can be classified as chemical, ionization, matrix, emission, spectral and background interferences. Chemical interference refers to a component that does not dissociate completely in the flame and can be solved by adding sequestering agent or by increasing the temperature of the flame. Ionization interference occurs when the flame temperature is too high and analyte produces ions. Adding ionization suppressor to the sample, standards and blanks can solve this problem. Matrix interferences may increase or decrease the signal. This occurs because the characteristics of the sample matrix (viscosity, burn speed, surface tension) may differ from the calibration curve standards. Standardized samples matrix should mimic sample matrix. Background interference can be solved by measuring its intensity and subtracting it from the total. Table 3.10 presents effects of interference and the correction needed for major element analysis.

Calibration Curves and Blanks

In theory, atomic absorption should follow Beer-Lambert's law with absorbance being directly proportional to concentration. However, departures from linearity are often encountered. A calibration curve that covers the range of concentrations found in the sample should be prepared. Any deviation of the standard from the original calibration curve can be used to correct the analytical result. The idealized calibration curve is stated by Beer's law that the absorbance of an absorbing analyte is proportional to its concentration.

Analytical blank (zeroing solution) consists in measuring the difference between the incident intensity source and the intensity transmitted and measured by the detector in a sample containing none of the analyte, usually a pure solvent (e.g., deionized water). In order to establish the calibration curve of a given element, standard solutions will be prepared by dilution to obtain a range of concentrations. Analytical interference in the determination of trace elements by AAS should be considered. Reagent blank solution is not to be confused with analytical blank, which is intended to detect possible sources of contamination from reagents, the work environment, or sample processing equipment.

Table 3.10 Major elements measured by AAS: effects of interference and their correction. Adapted from Table 4.6 (p. 125) in Potts [192]

Element	Line (nm)	Oxidizing gas	Interference	Action taken
Ca	422.7	Air	Si, Al, P Si, Al, $LiBO_2$, Ti Al, SO_4	Add La to sample and standard solutions[a] Add 1% La to sample and standard solutions; add Li to standard solutions; keep Si below 170 ppm[b] Use the nitrous oxide flame[c]
Ca	422.7	N_2O	Ionization	Add K to sample and standards solutions[a, c]
Na	589.0	Air	Alkalis $LiBO_2$ Ionization	Add Cs to sample and standard solutions[a] Add Li to standard solutions[b] Add excess K to samples and standard solutions[c]
K	766.5	Air	Alkalis Na, $LiBO_2$	Add Cs to samples and standard solutions[a] Add Li to standard solutions; add Na to standard solutions when required (i.e., when Na >5 ppm[b])

[a] According to Langmyhr and Paus [147]
[b] According to Van Loon and Parissis [251]
[c] According to Boar and Ingram [17]

Calibration and test sample solutions should exhibit absolutely identical behaviour in the atomizer. Interference can be compensated by matrix matching by mixing the interferent with the calibration solutions.

Limitations

According to Gupta and Gupta [99], the main limitations of AAS techniques are:

(1) Determination is limited to metals and metalloids.
(2) Time consuming and tedious sample preparation.
(3) Destructive analysis.
(4) Only one element measured at a time.

Atomic Absorption Spectroscopy with Graphite Furnace (GFAAS)

The principle of the GFAAS measurement is the same as flame AAS. In GFAAS, samples are deposited in a small graphite tube, which can then be heated by the electrothermal atomizer to vaporize and atomize the analyte. In GFAAS, a small volume of the sample solution, between 10 to 50 μL, is dispensed into the atomizer. The temperature is increased by steps to remove the solvent and concomitants before atomization. The entire aliquot introduce into the graphite tube is atomized rapidly (1 to 3 s) and generates a peak-shaped, time-dependent signal. The integrated absorbance is proportional to the mass of the analyte [262]. According to these authors, the sensitivity of GFAAS is two to three orders of magnitude higher than of FAAS because the entire sample is atomized and the residence time of the atoms in the absorption volume is much longer. The graphite furnace technique is also suitable for the direct analysis of solid samples. This is an important advantage considering complexity and problems related to every solubilisation procedure [262]. However, direct solid analysis presents also a disadvantage. In general, poor repeatability precision is obtained compared with solution analysis, which is largely due to the inhomogeneity of samples. For that reason, the solid sample should be ground to a particle size ≤50 μm for GFAAS analysis. Also see Sect. 3.2.1.3.

GFAAS is one of the most sensitive and robust techniques for trace element determination in liquid and in solid samples [262]. According to these authors, the typical working range for GFAAS is μg/L to ng/L in solution, and mg/kg to μg/kg in solid samples.

Flame Optical Emission Spectrometry (FOES)

The introduction of ICP-OES in the mid-1970s contributed to terminate the use of FOES except for a very few elements in a specific application. The FOES instrument is often called flame photometer. This low-resolution spectrometer is mainly used for the determination of the alkali metals Na, K, and Li, and occasionally the alkaline earth elements Ca and Ba [262] because these elements are sensitive to flame photometry due to their low excitation energies. FOES uses low-temperature flames with propane, butane, or natural gas as the fuel and air as the oxidant to avoid ionization of alkali elements. According to vlab.amrita.edu [253], the FOES technique has some advantages: it is a simple quantitative analytical method based on flame analysis, it is inexpensive; determination of elements such as alkali and alkaline earth metals is performed easily; quick, convenient, and selective and sensitive to even parts per million (ppm) to parts per billion (ppb) range. However, some disadvantages are linked to the FOES technique such as the difficulty to obtain accurate results of ions with higher concentrations, and the need for standard solutions.

3.3.4 Inductively-Coupled Plasma Spectrometer (ICP, AES (OES) and MS)

Inductively-Coupled Plasma - Optical Emission Spectrometry (ICP-OES)

ICP-OES (Inductively Coupled Plasma - Optical Emission Spectrometry, also referred to as Inductively Coupled Plasma - Atomic Emission Spectrometry) is an analytical technique used for the detection and quantitation of chemical elements. The development of the inductively coupled plasma source with an effective solution nebulizer revolutionized atomic emission-based atomic spectrometry [193]. The concise book by Dean [55] presents an accessible overview of the technique of inductively coupled plasma spectroscopy from the samples and standards preparation to the analysis to diverse applications.

ICP-OES is a multi-element technique that works by light excitation of specific lines of an element. The lower limit of detection for ICP-OES is parts per billion (ppb or ng/g). The method consists in spraying the solution in a plasma. Sample introduction via a nebulizer/spray chamber is key to the efficiency of atomization/ ionization in an ICP. The efficiency of the nebulization is low with less than 2% for a pneumatic nebulizer and the order of 10% for an ultrasonic nebulizer [55].

The ICP has sufficient thermal energy to ensure effective atomization of most analytes. The plasma is extremely hot with temperature close to 6000–7000 K and can reach up to 10 000 K in the induction zone. In the hot environment of an ICP, an atom is easily excited to an upper energy level. When the electrons drop back to ground state it is accompanied by the emission of a photon of radiation. The only elements that cannot be measured by these methods are H, C, O, N and halogens. As in all optical methods, the different wavelengths of the light beam are separated on an optical network, a device that makes it possible to sort the wavelengths. Emission spectra are complex and need a careful choice of analytical lines and backgrounds. A high-resolution monochromator is also needed to minimize spectral overlap interferences [193]. Inductively coupled plasma excitation works for almost all elements and many levels of energy per element. This method is relatively sensitive but the large number of lines requires solving interferences between the elements and minimizing background radiation emanating from the plasma gases and other elements. Spectral interference for atomic emission spectroscopy consists in spectral overlap and matrix effects [55]. Spectral overlaps are well known and can be avoid by the selection of alternative spectral emission line [252]. Matrix interferences come from the sample introduction process and is often due to the presence of easily ionizable elements such as alkali metals. The limiting precision of ICP-OES measurements is linked to noise in the signal and emission background. Potts [193] reported that the precision of major element determinations is not as good as XRF.

Advantages

- Up to 70 elements can be measured simultaneously in a single sample analysis.
- The useful working range is over several orders of magnitude.
- Instrumentation is suitable to automation, improving accuracy, precision and number of samples submitted for analysis.
- The combination of ICP-OES and ICP-MS is very powerful in determining an extremely wide range of elemental concentrations, while maintaining excellent accuracy and precision.

Limitations

- Only water-dissolved samples can be analysed (solids must be digested prior to analysis).
- Analysis can be complex with possibilities of spectral interference if the wavelength of the element of interest is very close to that of another element.
- Multi-charge or polyatomic ionic species can create quantification difficulties.

Inductively-Coupled Plasma - Mass Spectrometry (ICP-MS)

ICP-MS (Inductively Coupled Plasma Mass Spectrometry) is a well-established technique for very sensitive, trace and ultra-trace multi-element analysis of more than 70 elements. This method can also be used for determination of isotopic ratios. ICP-MS is more sensitive than flame AAS, GFAAS or ICP-OES. The greatest disadvantage of ICP-MS is the elemental isobaric interference caused by the different elements that form atomic ions of the same mass to charge ratio. ICP-MS, as for other spectral measurements, requires a set of well-defined standards and reference materials for accurate calibration. The following overview is based on the writings of Olesik [184], Bulska and Wagner [30], and Broekaert [21] and provides an introduction to this now widespread method of analysis.

The principle is based on the ionization of the atoms of the chemical element to be measured. Sample solutions are converted into liquid aerosols using a nebulizer. The aerosol passes through a spray chamber that removes large droplets and limits solvent load [184]. ICP produces the ionization of the solutions to be analysed by inductively coupling the nebulized solution in a stream of argon inside the high temperature plasma torch (5000 to 10000 K). The choice of argon is related to its high ionization energy. In fact, the argon plasma can efficiently ionize most elements in the periodic table. Thus, the element to be analysed is first atomized and then ionized. Moreover, argon is preferred because it is relatively inexpensive, inert, and monatomic.

Most ICP-MS instruments use a quadrupole mass spectrometer that rapidly, sequentially measures different masses. Detection limits are part per trillion (ppt, ng/L) or below for many elements with a linear dynamic range up to nine orders of magnitude [184]. A quadrupole is a simple mass spectrometer for the separation of the different mass to charge ratios in the ion beam wherein the trajectory of the ions injected between four cross-potential rods is disturbed by radio frequencies so that only the ions of a particular mass to charge ratio pass through the device to be counted on the detector. Such a device has no magnet and therefore allows a very fast scan of all the mass to charge ratios. In fact, for elemental analysis at concentrations of 1 part per billion (ppb, μg/L) or above measurement times of about 1 s per mass are typical, so that 60 elements could be scanned in about 1 min. However, in ICP-MS, the concentrations of dissolved solids in solutions are generally limited to 0.1–0.5% by weight because the sample is physically transported into the mass spectrometer and high concentrations of dissolved solids can cause different problems.

The ICP-MS allows determination of elements with atomic mass ranges from 7 to 250 (Li to U). However, element species with atomic masses in the vicinity of those of Ar, ArO, ArN (40–56) are unusable because they are saturated by the entrainment gas. In addition, certain species like e.g., chlorine (Cl) easily combine with Ar, interfering with heavier elements like arsenic (As). For the other masses, it remains only to compare the respective height of the signals for the sample and the standard. This method is very sensitive.

Table 3.11 Comparative detection limits of several analytical methods. Units are in μg/L (ng/mL, ppb). (From Thomas [240])

Element	Flame AAS	GFAAS	ICP-OES	ICP-MS
Ca	1.5	0.01	0.02	0.05
Na	0.3	0.005	0.5	0.0003
K	3	0.005	1	0.0002

Advantages

- High sensitivity with very low detection limits of most elements (in ppt to ppq concentrations).
- Simultaneous multi-element analysis possible.
- The ability to obtain isotopic information, especially with a multi-collector MS-instrument.

Limitations

- Substantially higher instrumentation cost than ICP-OES.
- Elements such as Ca and Fe are difficult to determine because of mass spectral interferences.
- Susceptible to instability with samples with high levels of total dissolved solids.

Table 3.11 presents comparative detection limits of different analytical techniques. According to Jenkins [129], the key distinction between AAS and ICP methods is the temperature during the absorption process. In AAS, the temperature of the air-acetylene flame is around 2400 K and that of NO_2-acetylene is around 3200 K. By comparison, the argon plasma torch (ICP) has a temperature of 10000 K and the temperature of the aerosol droplets of the sample is on the order of 6000 to 8000 K. A higher temperature generally allows for better dissociation, thus plasma source has an advantage over the flame.

3.3.5 Alternative Methods

Other methods include ion-selective electrodes (ISE). ISE is a membrane-based potentiometric device that is capable of accurately measuring the activity of ions in solution. Selectivity of these membranes for one ion over the other is determined by the nature and composition of the membrane materials used to make the electrode. Liquid and polymer membranes allow for the selective measurement of a wide variety of cations and anions (e.g., Na^+, K^+, Ca^{2+}, Ag^+, Cl^-, Br^-, and F^-). An ISE consists of a thin membrane across which only the intended ion can be transported. The transport of ions from a high concentration to a low one through a selective binding with some sites within the membrane creates a potential difference. The main limitations on ISE measurements are the effect of interference from other ions, the effect of the ionic strength of the solution reducing the measured activity relative to the true concentration at high concentrations and potential drift during a sequence of measurements [206]. Precision is rarely better than 1%.

3.4 In-Situ Determination of Alkali Contents by Electron Beam Techniques

3.4.1 Instrument Operating Principles

Since its invention by Von Ardenne [25], the scanning electron microscope - SEM has become a staple instrument as a powerful magnifier, as well as an apparatus for *in-situ* quantitative chemical analysis of individual mineral grains or parts thereof. Figure 3.13 below shows a modern SEM instrument with multiple detectors. The electron-probe micro-analyser - EPMA instrument was a further development of the SEM by Castaing [37], specifically (re-)designed for point-wise analysis. In-depth information on SEM instrumentation, micro-analysis and related subjects is collated in [96]. The successor edition [97] has a bit different setup and provides many hands-on instructions for practical use. Additional background for geological applications is given in [195], whereas the physical background is elaborated in detail in Reimer [196].

The SEM (or EPMA) instrument projects a thin electron beam onto the specimen, with a precisely known beam energy measured in kilovolts (kV), and a precisely known beam current measured in nano-Ampères (nA). Electrons can be regarded as

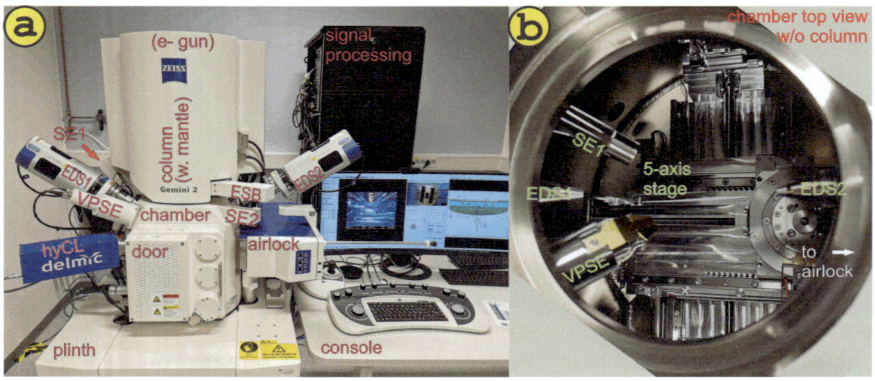

Fig. 3.13 **a** modern SEM instrument anno 2021, with various detectors: SE1 - secondary-electron; SE2 - in-lens SE; ESB - in-lens energy-sensitive backscatter; VPSE - low-vacuum SE; hyCL - hyperspectral cathodoluminescence; EDS1/2 - energy-dispersive (X-ray) spectrometer (these are only the Peltier-cooled electronics - the actual detector is inside). **b** top view inside chamber with column removed. The actual detectors ('X-ray sensors') are near the 'D' in 'EDS'

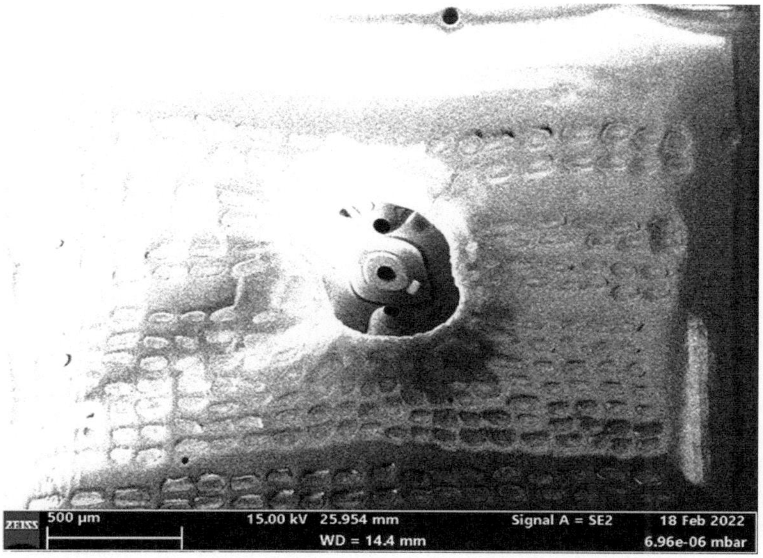

Fig. 3.14 Heavy charging of an electrically non-conducting (zircon) grain mount. Electrons bounced back by the charge build-up at the specimen surface image the pole-piece at the centre flanked by EDS detectors instead. Beam current was at 200 pA

negatively charged particles; therefore, the electron beam injects electrical charge into the sample material, which is led away from the point of beam incidence in conductive sample materials like metals and alloys. However, oxide and silicate minerals are generally poor electrical conductors so charge rapidly builds up at the specimen surface, deflecting further arriving electrons to actually reach the surface and interact with it in a predictable manner. In extreme cases, the deflected stray electrons show an image of the chamber interior rather than of the specimen (Fig. 3.14), obviously inhibiting clean observations with this type of detector (SE).

This detrimental analytical artefact is easily avoided by coating the specimen with a thin layer of electrically conductive material, which will be discussed further below. Alternatively, charging can be remediated by introducing gas (e.g., air, nitrogen, argon, helium) in the sample chamber, usually at a few tens of Pascals (Pa) which is then referred to as 'low-vacuum' or 'variable-pressure' SEM, in case of water vapor as a medium as 'environmental SEM' or ESEM (e.g., [138]). The beam electrons convert some gas molecules in their path into charged radicals which are very effective at dissipating the specimen's surface charge, but which also broaden the electron beam (aka. 'skirting') at the expense of lateral resolution for imaging and spot analysis [53]. Unless a given specimen cannot be coated (anticipating further analysis in instrumentation that is incompatible with such coating e.g., secondary-ion mass-spectrometry SIMS), a SEM instrument is preferably operated at high-vacuum (~10^{-6} Torr). With EPMA instruments, low-vacuum operation may not even be available.

3.4.2 Imaging with Secondary or Backscattered Electrons

Below the point of incidence on the specimen surface, the impinging electrons penetrate into the sample where they generate a variety of signals, e.g., heat, light, and various types of X-rays, at slightly different depths depending on the type of interaction with the material. We will now discuss a selection of electron-sample interactions, to better understand how these affect determination of alkali contents, and how to adjust instrument settings and operating conditions for optimal performance.

A fraction of the electrons in the beam directly hits an atomic nucleus in the specimen and rebound back elastically into the microscope chamber, keeping most of their initial energy. Thus, these so-called backscattered electrons (BSE) provide direct information on the relative atomic number Z of the sample material, which the BSE image shows as brightness differences-heavy phases appear bright, atomically lighter phases appear darker. An extreme example is given in Fig. 3.15. The image contrast is a result of the contrast in average molecular weight of both minerals, for uraninite ~90 (=[238 + 2 × 16]/3) versus quartz ~20 (=[28 + 2 × 16]/3). In more common samples, biotite and especially pyrite will stand out relatively brightly compared to biotite K-feldspar and quartz in granitic aggregate. In concrete, non-hydrated ferrite relics will shine brightly in a dark hydrated cement paste matrix. There is also inelastic scatter in which an electron

Fig. 3.15 Backscattered electrons (BSE) contrast, an example from the insoluble residue of a limestone, showing a round grain of heavy and therefore bright uraninite (UO_2) overgrown by a crystal of much lighter and thus darker quartz (SiO_2). Both minerals are of diagenetic origin

hits a light atomic nucleus which then effectively stops the electron. This 'stopping power' effect is inversely proportional to Z.

On their way into the sample material, beam electrons may alternatively hit an outer (loosely bonded) electron, knocking it out of its atomic orbit to emit from the specimen surface. These secondary electrons (SE) have much lower energy (~1–15 eV) than the incoming beam and are effectively collected by the positively charged (+300–400 V) cage of an SE detector. Because of their low energy, secondary electrons emit from small volume very close to the specimen surface (deeper SEs simply don't make it that far), and thus carry important information about specimen morphology. Therefore, SE imaging mode is most effective to inspect the surface of a polished section for preparation artefacts like scratches, pull-outs, undercutting, as well as dirt speckles, fingerprints, and the like. SE imaging is best suited to study e.g., fracture surfaces, crystal morphology, and other 'topographic features'. Both BSE and SE imaging are helpful in selecting locations suitable for *in-situ* analysis of chemical compositions. Modern SEM (and EPMA) instruments have both SE and BSE detectors, more advanced instruments often have several variants (e.g., in-chamber, in-column, low-vacuum) to enable additional imaging modes (see e.g., [97]) (Fig. 3.13).

3.4.3 Generation and Detection of X-rays for Analysis

Another fraction of the instrument's primary beam interacts with electrons close to the atomic nucleus, generating X-rays. The number of (positively charged) protons in the atomic nucleus determines the ionization energy of the (negatively charged) electrons orbiting the nucleus, arranged in shells with discrete energy levels, a bit like an onion (the vegetable), designated K for the innermost shell and further outward L, M, N, O and P. The very lightest elements only have K-shells (H, He, Li, Be, B, C), whereas P-shells only are present in the heaviest elements (e.g., Ra, Th, U). The heavier the element (i.e., the higher atomic number Z), the higher energies the shell electrons have. The number of (electrically neutral) neutrons in the nucleus can be ignored here.

Now, if a K-shell electron gets knocked out by an incoming beam electron, the K-vacancy is filled by a sibling-electron from the L-shell (or higher, subject to availability). The transition of the 'higher' L-electron into the 'lower' K-shell invokes release of excess energy, which is emitted as an X-ray photon with a specific energy characteristic for the emitting element species, accordingly termed 'characteristic X-ray'. These characteristic X-ray photons can be recorded, where the energy is a measure for the emitting element's identity, and the intensity (i.e., the number of X-rays per time-unit) a measure for its relative concentration. X-ray photons are not as easily absorbed or scattered by the sample matter or the conductive coating as electrically charged electrons (especially the relatively slow secondary ones), so that X-ray photons coming from greater depth may also escape the sample surface and be detected. Consequently, the actual focus point from which X-ray photons emergence is just below the specimen surface, which is relevant especially for WDS (see below).

It has been shown empirically that for efficient induction of characteristic X-rays, the beam energy needs to be twice as high as the ionization energy, in this context also known as 'critical excitation energy'. For iron Fe, this critical energy is at 7112 eV or 7.112 keV for the K-L transition [265], which is the key reason to operate a SEM or EPMA instrument at 15 kV beam acceleration voltage if the user has an interest to assess Fe in the sample material using its K-line. Yet heavier elements require higher acceleration voltages (up to a certain point, due to instrumental limitations), whereas lighter element species like e.g., K, Si only need <10 kV, and Na, Mg, Al, S even <5 kV. In reality, generation of X-rays by an electron beam is rather more complicated than sketched above - the interested reader is referred to Goldstein et al. [96, 97]) for further detail.

Many if not most modern SEM instruments are equipped with one energy-dispersive (X-ray) spectroscopy - EDS (or EDX) detector. Older detector types have a Dewar vessel containing liquid nitrogen to cool the signal amplifier and the X-ray sensor to −196 °C, in order to reduce electronic noise that otherwise might interfere with the signal to be detected. The part actually detecting the X-rays (the sensor) is a piece of lithium-doped silicon (referred to as "Si (Li)", pronounced 'silly') with 10 mm^2 surface area, the size of a tiny collar button on a dress shirt. Modern EDS detectors have smaller and faster low-noise amplifiers so that electrical Peltier-cooling is sufficient. Here, the sensor consists of a specially designed non-Li-doped semiconductor silicon wafer (hence: silicon-drift detector - SDD) with a much larger surface area accepting X-ray photons, (up to 180 mm^2 i.e., the size of a button on a clown suit, and still increasing).

EDS detectors are mounted to the instrument chamber on a port, with the cooling apparatus outside and the sensor + amplifier tandem on a stick (containing a 'heat pipe' to extract heat) close to the sample, 'watching' as it were the point of electron beam incidence. On a specimen without topography (i.e., a well-polished section), the X-ray photons are emitted in a hemisphere centred on the point of beam incidence. The closer an EDS detector is to the point of emission (i.e., the shorter its working distance in mm), the greater the part of all photons emitted it is able to catch. Modern SDD-type detectors have a double advantage compared to older Si (Li)-type: they have a shorter working distance, *and* they have a larger area. To get an idea, a 10 mm^2 Si (Li) at 15 mm working distance 'sees' ~0.14% of the entire X-ray emission-hemisphere (with 15 mm radius), compared with a modern 65 mm^2 SDD-type detector at 8 mm working distance which sees ~6.1% of its hemisphere (with 8 mm radius). All else (of instrument settings) the same, the SDD will receive well over forty times more X-ray photons than the Si (Li) in this example.

When an X-ray photon enters the detector, it generates electrons in the sensor body proportional to its energy, a few hundred for each X-ray quantum-keV. This extremely small charge (~10^{-17} C) is converted to a voltage pulse by a pre-amplifier, which can then be processed and counted. X-ray photons below a certain threshold energy are indistinguishable from noise, photons of higher energy are collated and presented as a histogram of energy increments, overall representing a spectrum of say 0–10 keV in which peak positions can be assigned to specific element species, and the area under each peak is a measure for its relative abundance.

The generated electrons from the sensor are multiplied by the amplifier located immediately after, and the boosted signal is processed by the detector's hardware to remove artefacts. This takes a little bit of time during which incoming X-rays are not recorded - this is referred to as 'deadtime', usually as a percentage. Fast processing shortens deadtime but at the expense of accuracy, whereas slow processing improves accuracy of the detector output with fewer signal artefacts, but which also increases deadtime and the amount of X-ray photons not getting counted and thus being excluded from the analysis. The instrument operator needs to find an agreeable balance between these extremes. Detection of X-rays by EDS detectors and the intricacies of in-detector signal processing are elaborated in great detail in Goldstein et al. [96, 97] and Reimer [196]. EDS detectors are normally set to a fixed working distance upon installation and have in principle no moving parts. They are robust and real workhorses, and their convenient size allows mounting several detectors on a single SEM instrument.

Before EDS detectors even existed, X-ray photons were counted by a detector type with a different working principle, known as the wavelength-dispersive spectrometer - WDS. The wavelength (λ) of electro-magnetic radiation is inversely related to its energy (E) by the equation $E = h \cdot [c/\lambda]$, where h = Planck constant, and c = the speed of light. Thus, EDS and WDS spectrometers are both really looking at the same phenomenon but from a different perspective. It also means that the energy of any given characteristic X-ray photon can be presented as a wavelength instead. In contrast to an EDS detector which records a complete X-ray spectrum, a WDS detector uses the Bragg condition (i.e., the principle equation for X-ray diffraction) to diffract incoming X-rays on a curved crystal plane to focus on a single characteristic wavelength peak. This requires complex mechanics with precision parts that only fit outside the sample chamber in a separate housing kept at chamber vacuum to minimize scatter. The point of X-ray emission is 'watched' via the WDS port in the SEM chamber wall, and the sensor may in certain settings be as much as a few hundred millimetres away from the point where the X-rays are emitted (measured along the beam path), in other cases close to the port at maybe half that distance. This means that a WDS thus receives only a very small fraction of X-ray photons, much smaller than a typical EDS detector. On the bright side of it, all photons belong to the selected peak only (plus its background) and are not shared by a full spectrum. X-rays are registered by a so-called proportional counter, consisting of a gas-filled tube with a central anode wire kept at an adequately high potential. Photons entering the sensor through a thin window ionize a few gas molecules that are then drawn to the anode, ionizing more gas molecules on the way causing an ionic avalanche proportional to the energy of the incoming X-ray, and the resulting current is then amplified and measured.

The sum outcome of this detector design is that accuracy and precision of analysis by WDS on an SEM are significantly higher compared to EDS, with a spectral resolution somewhere near 5–10 eV, versus 120–130 eV for EDS. Overlapping and/or adjacent peaks that EDS sees as one single broad peak or with a shoulder are normally well-discriminated by WDS (though exceptions still exist). On the other hand, counting efficiency of WDS is much lower due to the much smaller fraction of the total amount of X-ray photons generated that the detector at its far away position actually sees, so counting times are generally long. This can be compensated for by increasing beam current, but that may also exaggerate element migration or other beam exposure artefacts as discussed above. The discussions in Spray and Rae [221] and Morgan and London [172] are very helpful to define optimum settings for the sample material and instrumentation at hand.

Also, a WDS detector assesses only a small part of the whole spectrum, for one selected element. To compensate for that, EPMA instruments are commonly equipped with up to five WDS detectors, positioned upright around the sample chamber. Measuring ten element species then requires two runs of five species each, with some additional time to refocus the spectrometers in between. SEM instruments usually have only a single WDS mounted so that measuring multiple elements by WDS will take very long time and further increases the chance for beam exposure artefacts. In addition, inherent to its design with an internal X-ray focusing mechanism, a WDS detector is much more sensitive to even micrometer changes in specimen height (as well as surface roughness) from, e.g., refocusing than an EDS, especially vertically mounted WDS where both specimen and sensor movement are in the same plane. Inclining the detector's focusing plane (as more common on SEM instruments) at an angle to the electron beam makes it slightly less sensitive to specimen surface roughness, but is ineffective against specimen tilt and non-planarity [96].

3.4.4 Understanding Beam-Specimen Interaction and What Actually Is Analyzed

The acceleration voltage directly influences the penetration depth of electrons into the sample matter. The so-called 'interaction volume' that generates detectable signal below the point of incidence on the specimen surface is roughly droplet shaped. Wong and Buenfeld [265] calculated interaction depths and volumes for backscattered electrons for selected cement-based materials common in hydrated cement paste using Monte Carlo simulation, for different acceleration voltages. The interaction volume increases exponentially with acceleration voltage. Furthermore, it is important to understand that the reported values were modelled assuming contiguous phases, which however in real-life phases commonly have a grain size (much) smaller than the modelled interaction depth, and/or are arranged in a porous fabric with air (or epoxy if impregnated) between nano-sized mineral grains (e.g., C-S-H, ettringite).

The above bears consequences for the correct interpretation of analytical results, as the obtained compositional data relate to a volume considerably larger than the nominal beam diameter. It is therefore essential to assure in advance that a selected analytical spot is free of inclusions or underlying wedges of other phases, when measuring close to phase grain boundaries - neither SE or BSE imaging are able to reveal such features. Therefore, ordinary optical thin section petrography in transmitted (!) light remains an obligatory step to pre-select suitable locations for microanalysis. The same argument also means that results from the analysis of tiny 'crystal pompoms' or other minute details must be regarded with suitable skepticism.

In addition, the beam current is a crucial parameter to excite sufficient X-ray photons for reliable and swift analysis. Using formal definitions for SI-units of electrical charge (Coulomb - C) and current (Ampère - A), we can calculate that a commonly applied analytical beam current of 10 nA injects >60 billion electrons per second into the sample material. Only a small fraction of those electrons generates X-ray photons, and only a tiny fraction of those photons again reaches a detector where they are counted. Thus, most of the injected electrons are 'not detected' whether as SE or BSE, or the X-ray photons they may have generated. Regardless, these electrons certainly do affect the sample material in various additional ways. We will assume that electrical charging of the specimen surface is properly dissipated by a conductive coating, but in an insulating material like minerals are, that is little effective away from the surface [4]. It has been demonstrated that beam electrons may alter the oxidation status of iron in volcanic glasses [121], as well as cause sodium to migrate away reducing its apparent concentration (e.g., [127]), the effect increasing further with degree of hydration ([172], and references therein). Various strategies to negate beam-induced element migration have been published to serve target audiences working with different materials, e.g., Hunt et al. [124], Morgan and London [172], Hayward [110]. All recommend using a low beam current (2 nA), large beam diameter (20 μm), limited counting time, and assignment to a WDS detector (if available) at the beginning of an analytical run.

With above essentials about what an electron beam does to the material in a specimen and what signals the beam might generate (SE, BSE, X-rays), we can now discuss some properties of the relevant detectors to achieve the title of this section, i.e., to detect characteristic X-rays generated by alkalis Na and/or K in the sample material.

Because characteristic X-ray photons originate mostly from a volume below the point of beam incidence, they will pass through the sample material before they are emitted at the (conductively coated) specimen surface. While less prone than electrically charged electrons (beam, SE, BSE), X-rays certainly do interact with the sample matrix and the measured X-ray signal must be corrected for those effects. The correction algorithm consists of three main components, notably Z (for atomic number), A (for absorption), and F (for fluorescence). The atomic number Z has an effect on the amount of beam electrons that are subject to elastic and/or inelastic backscatter and are unable to contribute to generation of characteristic X-rays, which is corrected for in the Z-component of the algorithm. The A-component corrects for absorption - a characteristic X-ray generated by for example Na has the exactly right wavelength (=energy) to be completely absorbed by another Na atom on the way out of the matrix instead of being emitted for detection, leading to underestimation of Na. In turn, the F-component corrects for fluorescence - a characteristic X-ray is able to induce fluorescence in all element species with lower ionization energies (as actively used in XRF analysis- see Sects. 3.2.2.1 and 3.3.2 in this report), leading to overestimation of fluorescing species. Thus, ZAF matrix correction is necessary to compensate for these combined effects. Several variants exist, each with its own (dis-) advantages. Further detail is given in Goldstein et al. [97].

3.4.5 Polishing and Coating Are Essential—Not Optional

The focusing accuracy issues for WDS detectors mentioned above implicitly illustrate the importance of specimen preparation for *in-situ* determination of the chemical composition of minerals and related substances in cement and concrete, in particular the alkalis Na and K. Regardless of method, analytical results are only considered reliable if they (1) can be reproduced, *and* (2) 'the experiment' (i.e., the actual measurement) is repeatable. Measurements can be made repeatable by consistently applying rigorous analytical routine, including instrument settings and operation conditions that fit the sample at hand as well as to resolve the problem defined at the research outset, *and* to communicate these settings and conditions in a report or paper, in adequate detail. The key point of such rigorous approach is to replace randomness with control, which thus must comprise specimen preparation as a quintessential part of the analytical procedure. Therefore, the sample surface is de-randomized by thin sectioning and polishing. Routine procedures for thin sectioning and polishing or natural rocks are described in detail in Humphries [123], though Canada balsam for mounting has long been replaced by epoxy or acrylate alternatives.

Thin sectioning of damaged concrete, or otherwise fractured, deeply altered and/or poorly consolidated materials requires impregnation to mechanically support fragile and porous phases (e.g., C-S-H, ASR product, ettringite) through the cascade of abrasive steps to arrive at the desired section thickness and mirror polish [28]. Special care should be taken to ensure specimen flatness and planarity without scratches, undercutting of soft phases, or removal of fragile or incohesive constituents by plucking or poor adhesion with the mounting medium. To ensure reliable element mapping, the finished specimen also needs to be plane-parallel, meaning that the polished specimen surface is parallel to the carrier glass bottom surface, as opposed to wedged or curved. This ensures that when the stage is moved along x and y during element mapping, the section surface remains in focus which is especially important for WDS detectors (see above).

As already mentioned near the beginning of this section, oxide and silicate minerals are generally poor electrical conductors, and the charge delivered by the electron beam only very slowly dissipates away from the specimen surface (also see Fig. 3.14, and Bastin and Heijligers [4]). To minimize charge build-up, the specimen surface can be thinly coated with an electrical conductor like carbon or metal. For analytical purposes, coating with carbon is preferred because of its low atomic number ($Z = 6$). At low thickness to 10–20 nm, attenuation of incoming and outgoing (SE, BSE) electrons as well as of emitted X-ray photons is minimal [68, 133, 222], and correction algorithms can account for that. The transparent carbon coat is nearly neutral grey in colour and appears unobtrusive in optical petrography, and can swiftly be removed again by manual repolishing. A carbon coat is applied by thermal evaporation of a carbon filament or rod under (usually low) vacuum. Certain sample materials may be sensitive to high temperature paired with a bright flash used in some coating devices, other devices may have a gentler coating regime for carbon.

Noble metals coatings (e.g., gold, platinum, palladium and their alloys, iridium) are preferred for high-resolution microscopy due to their much more fine-grained nature that does not obliterate fine detail like carbon and to a lesser extent also gold (see e.g., [112, 226]). In addition, their high atomic number results in much stronger backscatter compared to carbon which is a benefit for microscopy *sensu stricto*, but a disadvantage for microanalysis due to an increased Z-component in the ZAF-matrix correction. Metals coats are applied by plasma sputtering, where the specimen is evacuated to high vacuum, and the plasma ignited using a strong magnetic field and argon gas as a carrier. Removal of noble metal coats is normally not that easy without harming the specimen, though gold may be removed using a simple solution [154]. A silver coat is also easy to remove [171] but will quickly tarnish upon exposure to air, like silver tableware does but faster. The same applies to many other metals that have been tried as conductive coating: rapid oxidation of extremely fine-grained metal coats exposed to ambient air has inhibited widespread application of e.g., nickel Ni and tungsten W for practical reasons [68, 112].

3.4.6 Evaluating Composition Data from EDS and WDS

The data from the detector software are raw element concentrations which are ZAF-corrected, and then converted to element oxides assuming stoichiometry, listed as weight percent (wt%). Iron represents the total amount of iron presented as FeO^T or $Fe_2O_3{}^T$, depending on the phase or phases analysed. For certain minerals/ groups, ferric and ferrous iron contents can be estimated by calculation cf. Droop [63]. The analytical sum total of all element oxides together should be within 100.00 ± 1.50 wt% for a calibrated EPMA instrument with WDS, for an SEM with EDS only perhaps within 100.00 ± 3.00 wt%. The presence of volatile oxides (e.g., CO_2, H_2O) in the nominal composition formula lowers the sum total as these are typically not analyzed. Thus, the analytical sum total serves as a simple check for the analytical quality. Normalization of the analytical total to 100.00 wt% destroys this effective opportunity see section 9.7.5 in Goldstein et al. [96], and sections 19.5 and 19.6 in Goldstein et al. [97].

As an example, consider calcite (nominally $CaCO_3$ or $CaO{\cdot}CO_2$ as oxides) consists of 40 wt% Ca which is the only species measured; the 12 wt% C and 48 wt% 3× O are not measured. The ZAF-corrected Ca-content is converted to the oxide CaO at 56 wt% which in this example is also the analytical sum total. If within say 56 ± 1.5 wt% then the result is acceptable, whereas anything outside (i.e., <54.5 wt% or >57.5 wt%) would be rejected. Normalization would present any results (i.e., too high, too low, or correct) as 100.00 wt%, thus eliminating this quality check.

Furthermore, optical thin section petrography is the most effective method for selection of locations suitable for *in-situ* microanalysis, with an immaculate surface free of scratches, no visible mineral or fluid inclusions, and with well-defined mineral/phase grain borders perpendicular to the section plane, to avoid underlying wedges that may compromise the result. Optical petrography is also the default method for mineral and other phases identification, essential for correct interpretation and evaluation of SEM–EDS and/or EPMA data. Nota bene: it is the combination of composition and crystal structure (as determined by their optical properties using thin section petrography) that identifies a mineral.

Another validation check is to recast weight percentages of element (oxides) into mineral composition formulae in **a**toms **p**er **f**ormula **u**nit - apfu. The protocol is explained with examples in the appendices to Deer et al. [56], prior knowledge of mineral/phase identity is essential (petrography!). Current mineral classification and nomenclature schemes are collated at http://cnmnc.units.it (accessed 20 November 2024).

A compilation of classification schemes for the most important mineral groups is given in Schertl et al. [210].

To conclude, even the most advanced instrumentation and most experienced operator cannot compensate for poor sample or specimen preparation, poor choice of analytical settings, poor understanding of the beam interactions with the sample material, and poor definition of the research target: what is it exactly that you want to know?

3.5 Reporting of Analytical Results, Analytical Error

As previously stated, accuracy and precision describe the reliability of an analytical measurement. Precision is considered as a measure of the repeatability that describes the degree of agreement between data obtained in exactly the same way while accuracy describes the correctness of analytical data [205, 269]. In analytical chemistry, the goal is to obtain data that are both accurate and precise.

An error can be defined as the difference between a measured value and the true value. Errors can be systematic or random. The former includes instrumental, personal and method errors [216]. In the case of a systematic error, increasing the sample size generally increases precision but does not improve accuracy. The accuracy is improved by eliminating the systematic error but does not change precision. The latter, the random errors, can be treated by statistical analysis based on repeated measurements. The measurement data can also be qualified according to their repeatability and reproducibility. Repeatability is the property of reproducing data under the same conditions, e.g., the same operator, using the same device, and the same method. On the other hand, the reproducibility is evaluated with different operators, using different instruments in different laboratories. Generally, the reliability of an analytical data increases with the number of measurements. The results should not be based on a single value but rather on the average value obtained from several measurements [263]. Analytical quality control techniques must be integrated into the routine operations. For example, analysing certified standard reference materials, blanks, duplicates, calibration standards, and other controls establish accuracy and precision.

Recorded analytical data should be reported using the right number of significant figures. Proper use of significant figures gives an indication of the reliability of the analytical method used. Only significant figures should be reported i.e., all figures known to be true and one last figure in doubt. For example, if a value is reported as 15.6 mg/L, the 15 must be firm values while the 0.6 is uncertain. Significant figures reflect the limit of the analytical method. For explanation and guidance, see e.g., Sections 17 in Petrucci et al. [189].

References

1. Andres, U., Bialecki, R.: Liberation of mineral constituents by high voltage pulses. Powder Technol. **48**(3), 269–277 (1986)
2. ASTM D75: Standard practice for sampling aggregates. American Society for Testing & Materials - ASTM, West Conshohocken/PA, p. 5 (2003)
3. Barneyback, R.S., Diamond, S.: Expression and analysis of pore fluids from hardened cement pastes and mortars. Cem. Concr. Res. **11**, 279–285 (1981)
4. Bastin, G.F., Heijligers, H.J.M.: Non-conductive specimens in the electron probe: a hitherto poorly discussed problem. In: Heinrich, K.F.J., Newbury, D.E. (eds.) Electron Probe Quantitation, pp. 163–175. Plenum Press, London/UK (1991)
5. Baumann, H.N.: Preparation of petrographic sections with bonded diamond wheels. Am. Miner. **42**(5–6), 416–421 (1957)
6. Beckhoff, B., Kanngießer, B., Langhoff, N., Wedell, R., Wolff, H.: Handbook of Practical X-Ray Fluorescence Analysis, p. 863. Springer, Berlin/DE (2006)
7. Berra, M., Costa, U., Mangialardi, T., Paolini, A.E.: Application of an innovative methodology to assessing the alkali-silica reaction in concrete. Mater. Struct. **48**(9), 2727–2740 (2014)
8. Bérubé, M.A., Pigeon, M., Dupont, N., Frenette, J., Langlois, M.: Expansion test methods for mass concrete exposed to alkali-aggregate reaction. Project CEA No.715 G 687, Can. Electr. Assoc. Montreal/CA, p. 99 (1992)
9. Bérubé, M.A., Pedneault, A., Frenette, J., Rivest, M.: Laboratory assessment of potential for future expansion and deterioration of concrete affected by ASR. CANMET/ACI Int. Workshop Alkali-Aggregate React. Conc. **270**(1), 267–271 (1995)
10. Bérubé, M.A., Duchesne, J., Rivest, M.: Alkali contribution by aggregates to concrete. In: Shayan, A. (ed.) Proceedings 10th ICAAR, Melbourne/AU, pp. 899–906 (1996)
11. Bérubé, M.A., Dorion, J.F., Rivest, M.: Distribution of alkalis in concrete structures affected by alkali-silica reactivity and contribution by aggregates. In: Bérubé, M.A., Fournier, B., Durand, B. (eds.) Proceedings 11th ICAAR, Québec/CA, pp. 139–148 (2000)
12. Bérubé, M.A., Frenette, J., Rivest, M., Vezina, D.: Measurement of the alkali content of concrete using hot-water extraction. Cem. Concr. Aggreg. **24**(1), 28–36 (2002)
13. Bérubé, M.A., Duchesne, J., Dorion, J.F., Rivest, M.: Laboratory assessment of alkali contribution by aggregates to concrete and application to concrete structures affected by alkali-silica reactivity. Cem. Concr. Res. **32**(8), 1215–1227 (2002)

14. Bérubé, M.A., Tremblay, C.: Chemistry of pore solution expressed under high pressure - influence of various parameters and comparison with the hot-water extraction method. In: Tang, M., Deng, M. (eds.) Proceedings 12th ICAAR, Beijing/CN, pp. 833–842 (2004)
15. Bérubé, M.A., Fournier, B.: Alkalis releasable by aggregates in concrete: significance and test methods. In: Tang, M., Deng, M. (eds.) Proceedings 12th ICAAR, Beijing/CN, pp. 17–30 (2004)
16. Bluhm, H., Frey, W., Giese, H., Hoppé, P., Schultheiß, C., Sträßner, R.: Application of pulsed HV discharges to material fragmentation and recycling. IEEE Trans. Dielectr. Electr. Insul. **7**(5), 625–636 (2000)
17. Boar, P.L., Ingram, L.K.: The comprehensive analysis of coal ash and silicate rocks by atomic-absorption spectrophotometry by a fusion technique. Analyst **95**(1127), 124–130 (1970)
18. Boer, R.H., Beukes, G.J., Meyer, F.M., Smith, C.B.: Fluoride precipitates in silicate wet-chemistry: implications on REE fractionation. Chem. Geol. **104**(1–4), 93–98 (1993)
19. Bouzahzah, H., Benzaazouah, M., Mermillod-Blondin, R., Pirard, E.: A novel procedure for polished section preparation avoiding internal particle settlement. Presentation at 12th International Congress Applied Mineralogy - ICAM, Istanbul/TR (2015)
20. BRE - Building Research Establishment: Alkali-Reaction in Concrete, Background to the guidance notes. Parts I–IV, BRE Digest 330, CRC Press Limited, London/UK (2004)
21. Broekaert, J.A.C.: Analytical Atomic Spectrometry with Flames and Plasmas, 2nd Edn., p. 413. Wiley, New York/USA (2006)
22. Broekmans, M.A.T.M.: Origin and mobility of alkalies in two Dutch ASR-concretes I: correlation with a damage rating index. In: Stamatakis, M., Georgali, B., Fragoulis, D., Toumbakari, E.E. (eds.) Proceedings 8th Euroseminar Microscopy Applied Building Materials - EMABM, Athens/GR, pp. 77–84 (2001)
23. Broekmans, M.A.T.M.: A simple and straight-forward damage rating index and its correlation with infiltrated chloride. In: Stamatakis, M., Georgali, B., Fragoulis, D., Toumbakari, E.E. (eds.) Proceedings 8th Euroseminar Microscopy Applied Building Materials - EMABM, Athens/GR, pp. 93–99 (2001)
24. Broekmans, M.A.T.M.: Sample representativity: effects of size and preparation on geochemical analysis. In: Fournier, B. (ed.) Marc-André Bérubé Symposium on Alkali-Aggregate Reactivity in Concrete. Proceedings 8th CANMET/ACI International Conference Recent Advances Concrete Technology, Montréal/CA, pp. 1–19 (2006)
25. Broekmans, M.A.T.M.: Petrography as an essential complementary method in forensic assessment of concrete deterioration: two case studies. Mater. Charact. **60**(7), 644–655 (2009)
26. Broekmans, M.A.T.M.: Deleterious reactions of aggregate with alkalis in concrete. In: Broekmans, M.A.T.M., Pöllmann, H. (eds.) Applied Mineralogy of Cement and Concrete. Rev. Mineral. Geochem. **74**(7), 279–364 (2012)
27. Broekmans, M.A.T.M., Fernandes, I.: Petrographic assessment of particulate materials: some statistical considerations. In: Çopuroğlu, O. (ed.) Proceedings 15th Euroseminar On Microscopy Applied to Building Materials - EMABM, Delft/NL, pp. 409–416 (2015)
28. Broekmans, M.A.T.M., Fernandes, I., Fredin, O., Margreth, A.: Polarization-fluorescence microscopy in the study of aggregates and concrete. Elements Mag. **48**(5), 321–326 (2022)
29. Buhrke, V.E., Jenkins, R., Smith, D.K.: A Practical Guide for the Preparation of Specimens for X-Ray Fluorescence and X-Ray Diffraction Analysis, p. 360. Wiley-VCH, New York/USA (2001)
30. Bulska, E., Wagner, B.: Quantitative aspects of inductively coupled plasma mass spectrometry. Philos. Trans. Roy. Soc. A **374**(2079), 20150369, p. 18 (2016)
31. Carmichael, R.S.: Practical Handbook of Physical Properties of Rocks and Minerals, p. 741. CRC Press, Boca Raton/FL (1989)
32. Carrasquillo, R.L., Snow, P.G.: Effect of fly ash on alkali-aggregate reaction in concrete. ACI Mater. J. **84**(4), 299–305 (1987)
33. Carrasquillo, R.L., Farbiarz, J.: Alkali-aggregate reaction in concrete containing fly ash - final report, Research Report 450–3F, Center for Transportation Research, University of Texas, Austin/TX, p. 136 (1989)

34. Casanova, I., Agulló, L., Aguado, A.: Aggregate expansivity due to sulfide oxidation - I. Reaction system and rate model. Cem. Concr. Res. **26**(7), 993–998 (1996)
35. CAS Common Chemistry: Solvent Yellow 43 (2024). https://commonchemistry.cas.org/detail?cas_rn=19125-99-6&search=19125-99-6. Accessed 25 Nov 2024
36. CAS Common Chemistry: Cuprate(1-), [C,C,C-tris[[[3-[(2-ethylhexyl)oxy]propyl]amino]sulfonyl]-29H,31H-phthalocyanine-C-sulfonato(3-)-κN^{29},κN^{30},κN^{31},κN^{32}]-, hydrogen, compd. with 3-[(2-ethylhexyl)oxy]-1-propanamine (1:1:1) (2024). https://commonchemistry.cas.org/detail?cas_rn=94277-77-7. Accessed 25 Nov 2024
37. Castaing, R.: Application des sondes électroniques à une méthode d'analyse ponctuelle chimique et cristallographique, Thèse de doctorat d'état, Université de Paris, Paris/FR. l'Office national d'études et de recherches aérospatiales, Publication (1952/55), p. 130 (1951)
38. Cazes, J.: Ewing's Analytical Instrumentation Handbook, 3rd Edn., p. 1037. Marcel Dekker, New York/USA (2005)
39. Cement and Concrete Association of Australia - CCAA: Alkali aggregate reaction: guidelines on minimising the risk of damage to concrete structures in Australia, CCAA Technical Report 47, Standards Association Australia, Handbook 79, p. 33 (1996)
40. Cement and Concrete Association of New Zeeland - CCANZ: Mackechnie, J.R., Freitag, S.A., Nicholson, C., Cook, D., Saggi, O., Hambling, J., Bennett, A.: TR3 Alkali Silica Reaction - Minimizing the Risk of Damage to Concrete, 3rd Edn., p. 75 (2021)
41. Chao, T.T.: Use of partial dissolution techniques in geochemical exploration. J. Geochem. Explor. **20**(2), 101–135 (1984)
42. Chao, T.T., Sanzolone, R.F.: Decomposition techniques. J. Geochem. Explor. **44**(1–3), 65–106 (1992)
43. Chappex, T., Scrivener, K.: Alkali fixation of C-S-H in blended cement pastes and its relation to alkali silica reaction. Cem. Concr. Res. **42**, 1049–1054 (2012)
44. Charlwood, R., Scrivener, K.L., Sims, I.: Recent developments in the management of chemical expansion of concrete in dams and hydro projects - Part 1: Existing structures. Presentation at: Hydro 2012 - Innovative approaches to global challenges, Session 13: Alkali-aggregate reaction in dams and hydro plants, Bilbao/ES, p. 17 (2012)
45. Claisse, F.: Accurate X-ray fluorescence analysis without internal standard. Quebec Department of Mines, Preliminary Reports, 927, pp. 24 (1956)
46. Concrete Society: Alkali silica reaction: minimising the risk of damage to concrete: guidance notes and model clauses for specifications, 3rd Edn. Concrete Society Technical Report 30, Sandhurst/UK, p. 72 (1999)
47. Cochran, M., King, A.G.: Two new types of holders used in grinding thin sections. Am. Miner. **42**(5–6), 422–425 (1957)
48. Constantiner, D., Diamond, S.: Pore solution analysis: are there pressure effects? In: Scrivener, K.L., Young, J.F. (eds.) Mechanisms of Chemical Degradation of Cement-Based Systems, pp. 22–29. CRC Press, London/UK (1997)
49. Constantiner, D., Diamond, S.: Alkali release from feldspars into pore solutions. Cem. Concr. Res. **33**(4), 549–554 (2003)
50. Cremer, M., Schlocker, J.: Lithium borate decomposition of rocks, minerals, and ores. Am. Miner. **61**, 318–321 (1976)
51. CSA A23.1-00/A23.2-00: Concrete materials and methods of concrete construction/Methods of test for concrete, Canadian Standards Association CSA, Ottawa/CA (2000)
52. CUR Recommendation 89-08: Measures to prevent concrete damage by the alkali-silica reaction. Official English translation, 2nd Edn., Centre for Civil Engineering Research and Codes, Gouda/NL, p. 48 (2008)
53. Danilatos, G.D.: Optimum beam transfer in the environmental scanning electron microscope. J. Microsc. **234**(1), 26–37 (2009)
54. Davies, G., Oberholster, R.E.: Alkali-silica reaction products and their development. Cem. Concr. Res. **18**(4), 621–635 (1988)
55. Dean, J.R.: Practical Inductively Coupled Plasma Spectroscopy, p. 197. Wiley, Chichester/UK (2005)

56. Deer, W.A., Howie, R.A., Wise, W.S., Zussman, J.: Rock-forming minerals 4B, Framework silicates: silica minerals feldspathoids and the zeolites, 2nd Edn., p. 982. Geological Society of London, London/UK (2004)
57. Delvigne, J.E.: Atlas of micromorphology of mineral alteration and weathering. Mineral. Assoc. Canada, Spec. Pub., 3, p. 495 (1998)
58. Diamond, S.: Effects of microsilica (silica fume) on pore-solution chemistry of cement pastes. Commun. Amer. Ceram. Soc. **66**(5), 82–84 (1983)
59. Dobbie, W.A., Burgess, C.J.: Discussion. Plastic-spray thin section cover. J. Sediment. Res. **48**(2), 671–674 (1978)
60. Doležal, J., Povondra, P., Šulcek, Z.: Decomposition Techniques in Inorganic Analysis, p. 224. Iliffe Books Ltd, London/UK (1968)
61. Drolet, C., Duchesne, J., Fournier, B.: Validation of the alkali contribution by aggregates to the concrete pore solution. Cem. Concr. Res. **98**, 10–23 (2017)
62. Drolet, C., Duchesne, J., Fournier, B.: Effect of alkali release by aggregates on alkali-silica reaction. Constr. Build. Mater. **157**, 263–276 (2017)
63. Droop, G.T.R.: A general equation for estimating Fe^{3+} concentrations in ferromagnesian silicates and oxides from microprobe analyses, using stoichiometric criteria. Mineral. Mag. **51**, 431–435 (1987)
64. DS 423.40: Testing of concrete - Hardened concrete - Production of fluorescence-impregnated thin sections (in Danish), Dansk Standardiseringsråd, Charlottenlund/DK, p. 10 (2002)
65. Duchesne, J., Bérubé, M.A.: Evaluation of the validity of the pore solution expression method from hardened cement pastes and mortars. Cem. Concr. Res. **24**(3), 456–462 (1994)
66. Durand, B.: A note about alkali contribution from aggregates in concrete affected by ASR. In: Bérubé, M.A., Fournier, B., Durand, B. (eds.) Proceedings 11th ICAAR, Québec/CA, pp. 169–177 (2000)
67. Duthaler, R., Weiß, S.: Mineralien reinigen und aufbewahren, 2nd Edn., p. 230. Christian Weise Verlag, München/DE (2008)
68. Echlin, P.: Handbook of Sample Preparation for Scanning Electron Microscopy and X-Ray Micro Analysis, p. 322. Springer, Heidelberg-Berlin/DE (2009)
69. Elsen, J., Lens., N., Aarre, T., Quenard, D., Smolej, V.: Determination of the w/c ratio of hardened cement paste and concrete samples on thin sections using automated image analysis techniques. Cem. Concr. Res. **25**, 827–834 (1995)
70. Engelhardt, H.J.: Zur Herstellung von Dünnschliffpräparaten von Evaporitgesteinen. Der Aufschluss **57**(1), 7–14 (2006)
71. Environmental Protection Agency - EPA: Flame atomic absorption spectrophotometry. Test Methods for Evaluating Solid Waste, Physical/Chemical Methods, 3rd Edn., EPA Publication SW-846, Method 7000B, 23 pp. (2007)
72. Feigl, F.: Qualitative analysis by spot tests. Inorganic and Organic Applications, 2nd English edition, translated from the 3rd German edition, p. 462. Elsevier Publishing Company, London/UK (1939)
73. Fernandes, I.: Caracterização petrográfica, química e física de agregados graníticos em betões. Estudo de casos de obra. Ph.D. thesis, Universidade do Porto, Porto/PT, p. 334 (2005)
74. Fernandes, I., Ribeiro, M.A., Couto, H., Santos Silva, A.: Alkali reactive aggregates: the importance of representative samples from the quarries. In: Proc.15th Euroseminar Microsc. Appl. Build. Mater. - EMABM. Delft University of Technology, Delft/NL, 207–214 (2015)
75. Fernandes, I., Ribeiro, M.A., Broekmans, M.A.T.M., Sims, I.: RILEM Recommended test method: detection of potential alkali-reactivity - AAR-1.2. Petrographic Atlas: Characterisation of Aggregates Regarding Potential Reactivity to Alkalis, p. 193. Springer, Berlin-Heidelberg/DE (2016)
76. Ferraz, A.R.: Estabilidade química de agregados de natureza granítica incorporados em betões de cimento Portland, p. 145. Unpublished Master Thesis, University of Lisbon (2017). http://hdl.handle.net/10451/30344. Accessed 25 Nov 2024

77. Ferraz, A.R., Fernandes, I., Soares, D., Santos Silva, A., Quinta-Ferreira, M.: Influence of the alteration of granitic rocks on alkalis release. In: World Multidisciplinary Earth Science Symposium - WMESS 2017, Prague/CZ, p. 9 (2017)
78. Filho, H.J.I., dos Santos Salazar, R.F., Capri, M.R., Neto, A.C., Kondracki de Alcântara, M.A., de Castro Peixoto, A.L.: State-of-the-art and trends in atomic adsorption spectrometry in atomic absorption spectroscopy. In: Farrukh, M.A. (ed.) Atomic Absorption Spectroscopy, pp. 13–36. InTech Open (2012). https://doi.org/10.5772/26076. Accessed 25 Nov 2024
79. Fletcher, W.K.: Analysis of soil samples. In: Fletcher, W.K. Hoffman S.J., Mehrtens M.B., Sinclair, A.J., Thomson, I. (eds.) Exploration Geochemistry: Design and Interpretation of Soil Surveys. Rev. Econ. Geol., **3**, 79–96 (1986)
80. Fournier, B., Bérubé, M.A.: Alkali–aggregate reaction in concrete: a review of basic concepts and engineering implications. Can. J. Civ. Eng. **27**, 167–191 (2000)
81. Fournier, B., Rogers, C.: Multi-laboratory study of accelerated mortar bar test and concrete prism expansion tests at 38° and 60°C. In: Broekmans, M.A.T.M., Wigum, B.J. (eds.) Proceedings 13th ICAAR, Trondheim/NO, pp. 647–656 (2008)
82. Fournier, B., Sanchez, L., Duchesne, J., Goyette, S.: Evaluation of the available alkali content in concrete through a modified hot-water extraction method. In: De Mayo Bernardes, H., Hasparyk, N.P. (eds.) Proceedings 15th ICAAR, São Paulo/BR, 10 pp. (2016)
83. French, W.J.: Comparison of the Canadian and British standard concrete prism tests and the effect of reduced permeability on test results. In: Poole, A.B. (ed.) Proceedings 9th ICAAR, London/UK, vol. 2, pp. 347–352 (1992)
84. Garcia, R., Baez, A.P.: Atomic Absorption Spectrometry (AAS), Atomic absorption spectroscopy. In: Farrukh, M.A. (ed.) Atomic Absorption Spectroscopy, pp. 1–12. InTech Open (2012). https://doi.org/10.5772/25925. Accessed 25 Nov 2024
85. Garrett, R.G.: Minor elements - the middle child between petrochemistry and geochemistry. Explore Newsl. **190**, 25–27 (2021)
86. Geikie, W.: The Founders of Geology, 2nd Edn., p. 486. MacMillan & Co. Ltd., London/UK (1905)
87. Giannini, E.R., Folliard, K.J., Zhu, J., Bayrak, O., Kreitman, K., Webb, Z., Hanson, B.: Non-destructive evaluation of in-service concrete structures affected by Alkali-Silica reaction (ASR) or delayed ettringite formation (DEF) - Final Report, Part I. No. FHWA/TX-13/0–6491–1, p. 228 (2013). http://library.ctr.utexas.edu/ctr-publications/0-6491-1.pdf. Accessed 25 Nov 2024
88. Gill, R. (ed.): Modern Analytical Geochemistry, Longman Geochemistry Series, p. 329. Routledge, Oxon/UK (1997)
89. Gillot, J.E., Rogers, C.A.: Alkali-aggregate reaction and internal release of alkalis. Mag. Concr. Res. **46**(167), 99–112 (1994)
90. Gillot, J.E., Rogers, C.A.: The behaviour of silicocarbonatite aggregates from the Montreal area. Cem. Concr. Res. **33**(4), 471–480 (2003)
91. Goguel, R., StJohn, D.A.: Chemical identification of Portland cements in New Zealand concretes. II. The Ba-Sr-Mn plot in cement identification and the effect of aggregates. Cem. Concr. Res. **23**, 283–293 (1993)
92. Goguel, R.: Alkali release by volcanic aggregates in concrete. Cem. Concr. Res. **25**, 841–852 (1995)
93. Goguel, R.: A new consecutive dissolution method for the analysis of slag cements. Cem. Concr. Aggreg. **17**, 84–91 (1995)
94. Goguel, R.: Selective dissolution techniques in AAR investigation: application to an example of failed concrete. In: Shayan, A. (ed.) Proceedings 10th ICAAR, Melbourne/AU, pp. 783–790 (1996)
95. Goguel, R., Milestone, N.B.: Alkali release from basalts and alkalinity of pore solution in mortar. In: Bérubé, M.A., Fournier, B., Durand, B. (eds.) Proceedings 11th ICAAR, Québec/CA, pp. 179–188 (2000)
96. Goldstein, J., Newbury, D.E., Joy, D.C., Lyman, C.E., Echlin, P., Lifshin, E., Sawyer, L., Michael, J.R.: Scanning electron microscopy and X-ray microanalysis, a text for biologists, materials scientists, and geologists, 3rd Edn., p. 689. Springer, Berlin-Heidelberg/DE (2003)

97. Goldstein, J., Newbury, D., Joy, D., Michael, J., Ritchie, N.W.M., Scott, J.H.: Scanning electron microscopy and X-ray microanalysis, a text for biologists, materials scientists, and geologists, 4th Edn., p. 550. Springer, Berlin-Heidelberg/DE (2018)
98. Graff, P.R.: Change in chemical composition of rock samples during preparation for analysis. Norges Geologiske Undersøkelse, NGU Bull. **300**, 27–31 (1973)
99. Gupta, A.P., Gupta, S.: Elemental profiling: its role and regulations. In: Farrukh, M.A. (ed.) Atomic Absorption Spectroscopy, 37–60. InTech Open (2012). https://doi.org/10.5772/25769. Accessed 25 Nov 2024
100. Gy, P.M.: Théorie et pratique de l'échantillonnage des matières morcelées, Editions PG, Cannes/FR, p. 595 (1975)
101. Gy, P.M.: Sampling of Particulate Materials, Theory and Practice, 2nd Edn. Elsevier Scientific Publishing Company, Amsterdam/NL, Developments in Geomathematics, **4**, p. 431 (1982)
102. Gy, P.M.: The analytical and economic importance of correctness in sampling. Anal. Chim. Acta **190**, 13–23 (1986)
103. Gy, P.M.: Sampling for Analytical Purposes, p. 153. Wiley, Chichester/UK/NY (1998)
104. Gy, P.M.: Part V: annotated literature compilation of Pierre Gy. Chemom. Intell. Lab. Syst. **74**, 61–70 (2004)
105. Haberlah, D.: SEM-EDS-based protocol for subsurface drilling mineral identification and petrological classification. In: Broekmans, M.A.T.M. (ed.) Proceedings 10th International Congress Applied Mineralogy - ICAM, pp. 265–273. Springer, Heidelberg, Berlin/DE (2012)
106. Haberlah, D.: Petrological reconstruction of the subsurface based on PDC-drill cuttings: an advanced rock typing approach. In: Broekmans, M.A.T.M. (ed.) Proceedings 10th International Congress Applied Mineralogy - ICAM, pp. 275–283. Springer, Heidelberg-Berlin/DE (2012)
107. Hansen, W.C.: Studies relating to the mechanism by which the alkali–silica reaction proceeds in concrete. J. Am. Concr. Inst. **15**, 213–227 (1944)
108. Hansen, W.C.: Release of alkalis by sands and admixtures in Portland cement mortars. ASTM Bull. **236**, 35–38 (1959)
109. Haugen, M., Skjølsvold, O., Lindgård, J., Østnor, T., Wigum, B.J.: Experience from petrographic analyses on aggregate separated from concrete cores. In: Broekmans, M.A.T.M., Jensen, V., Brattli, B. (eds.) Proceedings 9th Euroseminar Microscopy Applied Building Materials - EMABM, Trondheim/NO, p. 8 (2003)
110. Hayward, C.: High spatial resolution electron probe microanalysis of tephras and melt inclusions without beam-induced chemical modification. Holocene **22**, 119–125 (2011)
111. Hens, H.S.L.C.: Building Physics - Heat, Air and Moisture: Fundamentals and Engineering Methods with Examples and Exercises, 2nd Edn., p. 316. Wiley-VCH Verlag, München/DE (2012)
112. Heu, R., Shahbazmohamadi, S., Yorston, J., Capeder, P.: Target material selection for sputter coating of SEM samples. Microsc. Today **27**(4), 32–36 (2019)
113. Hickson, C.J., Juras, S.J.: Sample contamination by grinding. Can. Mineral. **24**(3), 585–589 (1986)
114. Hill, S.J.: Atomic absorption spectrometry: flame. In: Worsfold, P., Poole, C., Townshend, A, Miró, M. (eds.) Encyclopedia of Analytical Science, 2nd Edn., pp. 163–170. Elsevier (2005)
115. Hillebrand, W.F.: The analysis of silicate and carbonate rocks. United States Geological Survey - USGS. Bulletin **305**, 200 (1907)
116. Hobbs, D.W.: Effects of mineral and chemical admixtures on alkali-aggregate reaction. In: Okada, K., Nishibayashi, S., Kawamura, M. (eds.) Proceedings 8th ICAAR, Kyoto/JP, pp. 173–186 (1989)
117. Holmes, R.J.: Sampling mineral commodities - the good, the bad, and the ugly. J. South Afr. Inst. Min. Metall. **110**(6), 269–276 (2010)
118. Howarth, R.J.: Improved estimators of uncertainty in proportions, point-counting, and pass-fail test results. Am. J. Sci. **298**, 594–607 (1998)
119. Howarth, R.J., French, W.J.: A statistical study of aggregate testing data with respect to engineering judgment. In: Latham, J.P. (ed.) Advances in Aggregates and Armourstone Evaluation,

Geological Society of London, London/UK. Eng. Geol. Spec. Publication, **13**, pp. 169–183 (1998)

120. Howarth, R.J.: Estimation of the mean and reproducibility of aggregate compositions based on constituent mass data. Q. J. Eng. Geol. Hydrogeol. **33**, 247–256 (2000)
121. Hughes, E.C., Buse, B., Kearns, S.L., Brooker, R., Di Genova, D., Kilgour, G., Mader, H., Blundy, J.: High spatial resolution analysis of the iron oxidation state in silicate glasses using the electron probe. Am. Miner. **103**, 1473–1486 (2019)
122. Huff, D.: How to Lie with Statistics, p. 142. W.W. Norton & Company Inc., New York/USA (1954)
123. Humphries, D.W.: The preparation of thin sections of rocks minerals and ceramics. Oxford University Press, Kidlington/UK, Roy. Microsc. Soc., Handbook, **24**, p. 83 (1992)
124. Hunt, J.B., Clift, P.D., Lacasse, C., Vallier, T.L., Werner, R.: 7. Interlaboratory comparison of electron probe microanalysis of glass geochemistry. In: Saunders, A.D., Larsen, H.C., Wise, S.W. (eds.) Proceedings Ocean Drilling Program, Sci. Results, **152**, pp. 85–91 (1998)
125. Injuk, J., Van Grieken, R., Blank, A., Eksperiandova, L., Buhrke, V.: Chapter 6: Specimen preparation. In: Beckhoff, B., Kanngießer, B., Langhoff, N., Wedell, R., Wolff, H. (eds.) Handbook of Practical X-Ray Fluorescence Analysis, pp. 411–432. Springer, Berlin-Heidelberg/DE (2006)
126. Jakobsen, U.H., Laugesen, P., Thaulow, N.: Determination of water-cement ratio in hardened concrete by optical fluorescence microscopy. Am. Concr. Inst. ACI Int. Spec. Proc. **191**, 27–41 (2000)
127. Jbara, O., Cazaux, J., Trebbia, P.: Sodium diffusion in glasses during electron irradiation. J. Appl. Phys. **78**, 868–875 (1995)
128. JCAS I-31: Quantitative determination of quartz (phosphoric acid method) (in Japanese). Japan Cement Association Standard, JCAS 1-31, p. 11 (1996)
129. Jenkins, R.: X-Ray Fluorescence Spectrometry, 2nd Edn., p. 232. Wiley, New York/USA (1999)
130. Jochum, K.P., Nohl, U., Herwig, K., Lammel, E., Hoffmann, A.W.: GeoReM: a new geochemical database for reference materials and isotopic standards. Geostand. Geoanal. Res. **29**(3), 333–338 (2007)
131. Kagimoto, H., Sato, M., Kawamura, M.: Evaluation of degree of ASR deterioration in concrete and analysis of pore solutions. Concr. Lib. Japan. Soc. Concr. Eng. - JSCE, **36**, 281–293 (2000)
132. Kawamura, M., Koike, M., Nakano, K.: Release of alkalis from reactive andesite aggregates and fly ashes into pore solution in mortars. In: Okada, K., Nishibayashi, S., Kawamura, M. (eds.) Proceedings 8th ICAAR, Kyoto/JP, pp. 271–278 (1989)
133. Kerrick, D.M., Eminhizer, L.B., Villaume, J.F.: The role of carbon film thickness in electron microprobe analysis. Am. Miner. **58**, 920–925 (1973)
134. Kjellsen, K.O., Monsøy, A., Isachsen, K., Detwiler, R.J.: Preparation of flat-polished specimens for SEM-backscattered electron imaging and X-ray micro-analysis - importance of epoxy impregnation. Cem. Concr. Res. **33**, 611–616 (2003)
135. King, R.J.: The care of minerals. Part 1. The cleaning of minerals. J. Russell Soc. **1**(1), 42–48 (1982)
136. King, R.J.: The care of minerals. Part 2. The development of minerals. J. Russell Soc. **1**(2), 54–77 (1983)
137. Kingston, H.M., Jassie, L.B.: Introduction to microwave sample preparation: theory and practice. Anal. Chem. **61**(5), 300 (1988)
138. Kloprogge, J.T., Boström, T.E., Weier, M.L.: In situ observation of the thermal decomposition of weddellite by heating stage environmental scanning electron microscopy. Am. Miner. **89**, 245–248 (2004)
139. Knudsen, T., Thaulow, N.: Quantitative microanalyses of alkali–silica gel in concrete. Cem. Concr. Res. **5**, 443–454 (1979)
140. Kollek, J.J., Varma, S.P., Zaris, C.: Measurement of OH^- concentrations of pore fluids and expansion due to alkali–silica reaction in composite cement mortars. In: Proceedings 8th International Congress on the Chemistry of Cement—ICCC, Rio de Janeiro/BR, pp. 183–189 (1986)

141. Kratochvil, B., Mamba, S.: Microwave acid digestion of soil samples for elemental analysis. Can. J. Chem. **68**, 360–362 (1989)
142. Kreshkov, A.P., Chivikova, A.N., Zagorovskava, A.A.: A rapid method for determining free amorphous silica in clay. J. Anal. Chem. USSR **20**, 1295–1296 (1965)
143. Kwitko-Ribeiro, R.: New sample preparation developments to minimize mineral segregation in process mineralogy. In: Broekmans, M.A.T.M. (ed.) Proceedings 10th International Congress Applied Mineralogy - ICAM, Trondheim/NO, pp. 411–417. Springer, Berlin-Heidelberg/DE (2012)
144. Lamble, K.J., Hill, S.J.: Microwave digestion procedures for environmental matrices. Analyst **123**, 103R–133R (1998)
145. Lamé, F.P.J., Defize, P.R.: Sampling of contaminated soil: sampling error in relation to sample size and segregation. Environ. Sci. Technol. **27**, 2035–2044 (1993)
146. Langmyhr, F.J., Kringstad, K.: An investigation of the composition of the precipitates formed by the decomposition of silicate rocks in 38–40% hydrofluoric acid. Anal. Chim. Acta **35**(2), 131–135 (1966)
147. Langmyhr, F.J., Paus, P.E.: The analysis of inorganic siliceous materials by atomic absorption spectrophotometry and the hydrofluoric acid decomposition technique: Part 1. The analysis of silicate rocks. Anal. Chim. Acta **43**, 397–408 (1968)
148. Lastra, R., Cabri, L.J., Weiblen, P.W.: Comparative liberation study by image analysis of Merensky Reef samples comminuted by electric-pulse disaggregation and by conventional crusher. In: Lorenzen, L., Bradshaw, D.J. (eds.) Proceedings 22nd International Mineral Processing Congress, Cape Town/RSA, pp. 251–260 (2003)
149. Lastra, R., Paktunc, D.: An estimation of the variability in automated quantitative mineralogy measurements through inter-laboratory testing. Miner. Eng. **95**, 138–145 (2016)
150. Lastra, R., Petruk, W.: Mineralogical characterization of sieved and un-sieved samples. J. Mineral. Mater. Charact. Eng. **2**, 40–48 (2014)
151. LCPC: Essai de granulats. Détermination des alcalins solubles dans l'eau de chaux. Méthod d'essai No. 37, Laboratoire Central des Ponts et Chaussées, Paris/FR, p. 12 (1993)
152. Leng, Y.: Materials characterization. Introduction to Microscopy and Spectroscopic Methods, p. 337. Wiley-VCH, Berlin/DE (2008)
153. Le Roux, A., Ouali, A., Gu, J.: Alcali-réaction - Importance de la connaissance pétrographique dans la démarche préventive. Bulletin Des Laboratoires Des Ponts et Chaussées, Paris/FR **211**, 133–139 (1997)
154. Leslie, S.A., Mitchell, J.C.: Removing gold coating from SEM samples. Palaeontology **50**(6), 1459–1461 (2007)
155. Limandri, S.P., Carreras, A.C., Trincavelli, J.C.: Effects of the carbon coating and the surface oxide layer in electron probe micro analysis. Microsc. Microanal. **16**, 583–593 (2010)
156. Linge, K.L., Jarvis, K.E.: Quadrupole ICP-MS: introduction to instrumentation, measurement techniques and analytical capabilities. Geostand. Geoanal. Res. **33**, 445–467 (2009)
157. Locati, F., Marfil, S., Baldo, E., Maize, P.: Na_2O, K_2O, SiO_2 and Al_2O_3 release from potassic and calcic-sodic feldspars into alkaline solutions. Cem. Concr. Res. **40**, 1189–1196 (2010)
158. Lof, P.: Elseviers' Mineral and Rock Table. Wall chart, 0.71 × 1.35 m. Elsevier Publishers, Amsterdam/NL (1982)
159. Lombard, A., Youlton, B.: Quality control: QEMSCAN from a different angle 21st Generation Meeting International Minerals Association - IMA, Johannesburg/SA, Abstracts Volume, 310 (2014)
160. Longuet, P., Burglen, L., Zelwer, A.: La phase liquide du ciment hydraté. Revue Matériaux Constr. **676**, 35–41 (1973)
161. Loubser, M.: Chemical and physical aspects of lithium borate fusion. Master Thesis, University of Pretoria, Pretoria/ZA, p. 139 (2009)
162. Loubser, M.: Representative sampling of heterogeneous and complex materials - from lot to aliquot. Presentation at: Building Materials Analysis—Symposium III, Halle/DE (2013)

163. Lu, D., Zhou, X., Xu, Z., Lan, X., Tang, M., Fournier, B.: Evaluation of laboratory test method for determining the potential alkali contribution from aggregate and the ASR safety of the Three-Gorges dam concrete. Cem. Concr. Res. **36**, 1157–1165 (2006)
164. Margui, E., Van Grieken, R.: X-Ray Fluorescence Spectrometry and Related Techniques: An Introduction, p. 162. Momentum Press, New York/USA (2013)
165. Menéndez, E., Prendes, N., Márquez, C., Aldea, B.: Evaluation of granitic aggregates behaviour in relation with the alkaline extraction and compositional change in their phases. In: Drimalas, T., Ideker, J.H., Fournier, B. (eds.) Proceedings 14th ICAAR, Austin/TX, p. 10 2012)
166. Menéndez, E., García-Rovés, R., Aldea, B., Ruíz, S.: Alkali release of aggregates. Effectiveness of different solutions and conditions of test. In: De Mayo Bernardes, H., Hasparyk, N.P. (eds.) Proceedings 15th ICAAR, São Paulo/BR, p. 10 (2016)
167. Menéndez, E., García-Roves, R., Ruiz, S.: Alkali release from aggregates: contribution on ASR. Constr. Mater. **169**(CM4), 206–214 (2016)
168. Menéndez, E.: Estrategia integral de prevención de la reacción árido-álcali (Integral strategy of prevention of the alkali-aggregate reaction), Consejo Superior de Investigaciones Científicas (CSIC), Monografías del Instituto Eduardo Torroja, **430** (2019). http://hdl.handle.net/10261/211145. Accessed 25 Nov 2024
169. Menéndez, E., Santos Silva, A., Fernandes, I., Duchesne, J., Berra, M., De Weerdt, K., Salem, Y., Garcia-Rovés, R., Soares, D., Fournier, B., Mangialardi, T., Lindgård, J.: RILEM TC 258-AAA Round Robin Test: Alkali release from aggregates and petrographic analysis. Critical review of the test method AAR-8. Mater. Construcc. **72**(346), e279, p. 10 (2022)
170. Mills, W.J., Hamilton, R.D.: Chemical methods for analysis of silica: a critical literature review. Analytica Chimica Acta (286/1):3–7 (1994)
171. Mills, A.A.: Silver as a removable coating for scanning electron microscopy. Scanning Microsc. **2**, 1265–1272 (1988)
172. Morgan, G.B., London, D.: Effect of current density on the electron microprobe analysis of alkali aluminosilicate glasses. Am. Mineral. **90**, 1131–1138 (2005)
173. Mohs, F.: Dritter Abschnitt. Zweites Kapitel. II. Die Härte. §208. Grade der Härte, und das Verfahren, sie zu bestimmen. In: Grund-Riß der Mineralogie, Erster Theil, Terminologie, Systematik, Nomenklatur, Charakteristik, Mit 5 Kupfertafeln, In der Arnoldischen Buchhandlung, Dresden/DE, pp. 374–382 (1822)
174. Moussa, M.T.: Plastic spray thin section cover. J. Sediment. Res. **46**(1), 252–253 (1976)
175. Moussa, M.T.: Reply. Plastic spray thin section cover. J. Sediment. Res. **48**(2), 671–674 (1978)
176. Nakayama, K., Nakamura, T.: X-ray fluorescence spectroscopy for geochemistry. In: McDonough, W.F. (ed.) Analytical Geochemistry/Inorganic Instrumental Analysis, 2nd Edn., Treatise on Geochemistry, **15**(10), 181–194 (2014)
177. Ney, P.: Gesteinsaufbereitung im Labor, Ferdinand Enke Verlag, Stuttgart/DE, p. 157 (1986)
178. NF P18–454: Réactivité d´une formule de béton vis-à-vis de l´alcali-réaction. Essai de performance, Editions AFNOR, Paris/FR, p. 27 (2004)
179. Nicol, S.: On a method of so far increasing the divergency of the two rays in calcareous-spar, that only one image may be seen at a time. Edinburgh New Philos. J. **6**, 83–84 (1828)
180. Nixon, P.J., Page, C.L.: Pore solution chemistry and alkali aggregate reaction. Am. Concr. Inst. - ACI. Symp. Pap. **100**, 1833–1862 (1987)
181. Nixon, P.J., Collins, R.J., Rayment, P.L.: The concentration of alkalies by moisture migration in concrete - a factor influencing alkali aggregate reaction. Cem. Concr. Res. **9**(4), 417–423 (1979)
182. Norrish, K., Hutton, J.T.: An accurate X-ray spectrographic method for the analysis of a wide range of geological samples. Geochim. Cosmochim. Acta **32**, 431–453 (1969)
183. Obenauf, R.H., Nash, C., Martin, J., Bostwick, R., DeStefano, M., Akers, J., Tucker, K.: SPEX Handbook of Sample Preparation and Handling, 9th Edn., SPEX Sample Preparation, Metuchen/NJ, 152 pp. (2005)
184. Olesik, J.W.: Inductively coupled plasma mass spectrometers. In: McDonough, W.F. (ed.) Analytical Geochemistry/Inorganic Instrumental Analysis, 2nd Edn., Treatise on Geochemistry, **15**(17), 309–336 (2014)

185. Ouali, A.: Étude expérimentale du lessivage des alcalins et du silicium extractibles des granulats naturels. Thèse de doctorat, Université P. et M. Curie, Paris/FR, p. 176 (1997)
186. Parmelee, C.W., Monack, A.J.: The solubility of feldspars in water. J. Am. Ceram. Soc. **13**(6), 386–392 (1930)
187. Pearce, J.M.: Open-Source Lab - How to Build Your Own Hardware and Reduce Research Costs. Elsevier (2014). https://doi.org/10.1016/C2012-0-07249-3. Accessed 25 Nov 2024
188. Petersen, L., Esbensen, K.H.: Representative process sampling for reliable data analysis - a tutorial. J. Chemom. **19**, 625–647 (2005)
189. Petrucci, R.H., Herring, F.G., Madura, J.D., Bissonnette, C.: General Chemistry: Principles and Modern Applications, 4th Edn., p 1326. Pearson, Toronto/CAN (2017)
190. Pitard, F.F.: Pierre Gy's Sampling Theory and Sampling Practice, Heterogeneity, Sampling Correctness, and Statistical Process Control, 2nd Edn., p. 488. CRC Press, Boca Raton/FL (1993)
191. Plusquellec, G., Geiker, M.R., Lindgard, J., Duchesne, J., Fournier, B., De Weerdt, K.: Determination of the pH and the free alkali metal content in the pore solution of concrete: Review and experimental comparison. Cem. Concr. Res. **96**, 13–26 (2017)
192. Potts, P.J.: A Handbook of Silicate Rock Analysis, 662 pp. Springer Science+Business Media LLC, New York/USA (1992)
193. Potts, P.J.: Principles of atomic spectroscopy. In: McDonough, W.F. (ed.) Analytical Geochemistry/Inorganic Instrumental Analysis, 2nd Edn., Treatise on Geochemistry, **15**(09), 171–180 (2014)
194. Rajabipour, F., Giannini, E., Dunant, C., Ideker, J.H., Thomas, M.D.A.: Alkali–silica reaction: current understanding of the reaction mechanisms and the knowledge gaps. Cem. Concr. Res. **76**, 130–146 (2015)
195. Reed, S.J.B.: Electron Microprobe Analysis and Scanning Electron Microscopy in Geology, 2nd Edn., p. 206. Cambridge University Press, Cambridge/UK (2005)
196. Reimer, L.: Scanning Electron Microscopy Physics of Image Formation and Microanalysis, p. 527. Springer, Berlin, Heidelberg/DE (1998)
197. Reinicke, F.: III. Miscellen. Das Format des Objectträgers. Beitraege zur neuern Mikroskopie. Zweites Heft. Verlagsbuchhandlung von Rudolf Kunze, Dresden/DE, pp. 75–77 (1860)
198. RILEM Recommended Test Method: AAR-0 - Outline guide to the use of RILEM methods in assessments of alkali-reactivity potential of aggregates. In: Nixon, P.J., Sims, I. (eds.) RILEM Recommendations for the Prevention of Damage by Alkali-Aggregate Reactions in New Concrete Structures. Springer, Heidelberg, Berlin/DE. RILEM State-of-the-Art Report, **17**, 1–34 (2016)
199. RILEM Recommended Test Method: AAR-1.1 - Detection of potential alkali-reactivity - Part 1: petrographic examination method. In: Nixon, P.J., Sims, I. (eds.) RILEM Recommendations for the Prevention of Damage by Alkali-Aggregate Reactions in New Concrete Structures. Springer, Heidelberg-Berlin/DE. RILEM State-of-the-Art Report, **17**, 35–60 (2016)
200. Rivard, P., Bérubé, M.A., Ballivy, G., Ollivier, J.P.: Effect of drying–rewetting on the alkali concentration of the concrete pore solution. Cem. Concr. Res. **33**(6), 927–929 (2003)
201. Rodrigues, A., Duchesne, J., Fournier, B., Durand, B., Rivard, P., Shehata, M.: Mineralogical and chemical assessment of concrete damaged by the oxidation of sulfide-bearing aggregates: importance of thaumasite formation on reaction mechanisms. Cem. Concr. Res. **42**, 1336–1347 (2012)
202. Rogers, C.A., Hooton, R.D.: Leaching of alkalies in alkali-aggregate reaction testing. In: Okada, K., Nishibayashi, S., Kawamura, M. (eds.) Proceedings 8th ICAAR, Kyoto/JP, pp. 327–332 (1989)
203. Rogers, C.A., Hooton, R.D.: Reduction in mortar and concrete expansion with reactive aggregates due to alkali leaching. Cem. Concr. Aggreg. **13**(1), 42–49 (1991)
204. Rosiwal, A.: Vorläufige Mitteilung über eine neue Methode der Härtebestimmung durch Schleifen. Anzeige der Akademie der Wissenschaften in Wien **12**, 393 (1893)
205. Rousseau, R.M.: Detection limit and estimate of uncertainty of analytical XRF results. Rigaku J. **18**(2), 33–47 (2001)

206. Rundle, C.C.: A beginner's guide to ion-selective electrode measurements (2000). https://www.nico2000.net/Book/Guide1.html. Accessed 25 Nov 2024
207. Santos Silva, A., Fernandes, I., Ferraz, A.R., Soares, D.: Can certain alkali minerals explain the slow reactivity of granitic aggregates in dams? In: Sellier, A., Grimal, E., Multon, S., Bourdarot, E. (eds.) Swelling Concrete in Dams and Hydraulic Structures, pp. 93–105. Wiley-ISTE (2017)
208. Saini-Eidukat, B., Weiblen, P.W.: A new method of fossil preparation, using high-voltage electric pulses. Curator **39**, 139–144 (1996)
209. Schatz, P.: Tumbling Apparatus, US Patent **3,610,587**, p. 4 (1971)
210. Schertl, H.P., Mills, S.J., Maresch, W.V.: A compendium of IMA-approved mineral nomenclature. In: 22nd IMA General Meeting, Melbourne/AU. Schweizerbart'sche Verlagsbuchhandlung (Nägele u. Obermiller), Stuttgart/DE: p. 353 (2019)
211. Shackley, M.S.: X-ray Fluorescence Spectrometry (XRF) in Geoarchaeology, p. 245. Springer Science+Business Media LLC, New York/USA (2011)
212. Shayan, A., Wark, R.E., Moulds, A.: Diagnosis of AAR in Canning dam, characterization of the affected concrete and rehabilitation of the structure. In: Bérubé, M.A., Fournier, B., Durand, B. (eds.) Proceedings 11th ICAAR, Québec/CA, pp. 1383–1392 (2000)
213. Shehata, M.H., Thomas, M.D.A.: The effect of fly ash composition on the expansion of concrete due to alkali-silica reaction. Cem. Concr. Res. **30**, 1063–1072 (2000)
214. Sibilia, J.P.: A Guide to Materials Characterization and Chemical Analysis, 2nd Edn., p. 388. Wiley-VCH, Chichester/NY (1996)
215. Sims, I., Nixon, P.: Assessment of aggregates for alkali–aggregate reactivity potential: RILEM International Recommendations. In: Fournier, B. (ed.) Marc-André Bérubé Symposium on Alkali-Aggregate Reactivity in Concrete. Proceedings 8th CANMET/ACI International Conference Recent Advances Concrete Technology, Montréal/CA, pp. 71–91 (2006)
216. Skoog, D.A., Holler, F.J., Crouch, S.R.: Principles of Instrumental Analysis, 6th Edn., p. 1057. Thomson-Brooks Cole, Belmont/CA (2007)
217. Smith, D.K.: Particle statistics and whole-pattern methods in quantitative X-ray powder diffraction analysis. Powder Diffr. **16**(4), 186–191 (2001)
218. Soares, D., Santos Silva, A., Mirão, J., Ramos, V., Fernandes, I, Menéndez, E.: Assessment of alkalis released by aggregates. Contribution to the alkalinity increase and AAR development in concrete. In: Proceedings 2nd International Dam World Conference, LNEC, Lisbon/PT, p. 10 (2015)
219. Soares, D., Santos Silva, A., Mirão, J., Fernandes, I., Menéndez, E.: Study on the factors affecting alkalis release from aggregates into ASR. In: De Mayo Bernardes, H., Hasparyk, N.P. (eds.) Proceedings 15th ICAAR, São Paulo/BR, p. 10 (2016)
220. Spiers, C.J., Urai J.L., Lister G.S., Boland J.N., Zwart H.J.: The influence of fluid-rock interaction on the rheology of salt rock. Comm. Eur. Communities, Nucl. Sci Tech. Final Report EUR 10399 EN, p. 131 (1986)
221. Spray, J.G., Rae, D.A.: Quantitative electron-microprobe analysis of alkali silicate glasses: a review and user guide. Can. Mineral. **33**, 323–332 (2005)
222. Springer, G.: The role of carbon film thickness in electron microprobe analysis: a comment. Am. Mineral. **59**, 1121–1122 (1974)
223. Stark, D.: Alkali-silica reactivity - some considerations. Cem. Concr. Aggreg. **2**, 92–100 (1980)
224. Stark, D., Bhatty, M.S.Y.: Alkali-silica reactivity: effect of alkali in aggregate on expansion. Am. Soc. Test. Mater. - ASTM, Spec. Tech. Pub. **930**, 16–30 (1986)
225. St John, D.A., Goguel, R.L.: Pore solution/aggregate enhancement of alkalies in hardened concrete. In: Poole, A.B. (ed.) Proceedings of 9th ICAAR, London/UK, vol. 2, pp. 894–901 (1992)
226. Stokroos, I., Kalicharan, D., Van der Want, J.J.L., Jongebloed, W.L.: A comparative study of thin coatings of Au/Pd, Pt and Cr produced by magnetron sputtering for FE-SEM. J. Microsc. **189**(1), 79–89 (1998)

227. Stroeven, P., Guo, Z.: Sampling concrete by too small elements; what should we do to get reliable information? Image Anal. Stereol. **27**, 151–162 (2008)
228. Struble, L.J.: The influence of cement pore solution on alkali-silica reaction. Ph.D. Thesis, Purdue University, USA; Gaithersburg, Maryland: Report NBSIR 87–3632, National Bureau of Standards, US Department of Commerce, p. 265 (1987)
229. Stutzman, P.E.: Microscopy of clinker and hydraulic cements. In: Broekmans, M.A.T.M., Pöllmann, H. (eds.) Applied Mineralogy of Cement and Concrete, Rev. Mineral. Geochem. **74**(3), 101–145 (2012)
230. Šulcek, Z., Povondra, P.: Methods of Decomposition in Inorganic Analysis, p. 336. CRC Press Inc., Boca Raton/FL (1992)
231. Suzuki, T., Sensui, M.: Application of the microwave digestion method to the decomposition of rock samples. Anal. Chim. Acta **245**, 43–48 (1991)
232. Tagnit-Hamou, A., Saric-Coric, M., Rivard, P.: Internal deterioration of concrete by the oxidation of pyrrhotitic aggregates. Cem. Concr. Res. **35**(1), 99–107 (2005)
233. Talvitie, N.A.: Determination of quartz in the presence of silicates using phosphoric acid. Anal. Chem. **23**, 623–626 (1951)
234. Taylor, H.F.W.: Cement Chemistry, 2nd Edn., p. 459. Thomas Telford Publishing, London/UK (1997)
235. Taylor, H.E.: Inductively Coupled Plasma-Mass Spectrometry: Practices and Techniques, p. 294. Academic Press, San Diego (2000)
236. The University of Edinburgh: Sample preparation and sample requirements. School of Geosciences, Edinburgh/SCO (2021). https://geosciences.ed.ac.uk/about/facilities/all/ion probe/technical/samples/epoxy-resins. Accessed March 2025
237. Thomas, M.D.A.: The role of fly ash and slags alkalis in alkali-silica reactions in concrete. In: CANMET/ACI International Workshop on AAR in Concrete, Dartmouth, Nova Scotia, CA, pp. 181–204 (1995)
238. Thomas, M.D.A.: The role of calcium hydroxide in alkali recycling in concrete. In: Skalny, J., Gebauer, J., Odler, I. (eds.) Special Volume Calcium Hydroxide in Concrete, American Ceramic Society, Westerville, Mater. Sci. Concr. Series, **58**, 269–280 (2001)
239. Thomas, M.D.A.: The effect of supplementary cementing materials on alkali-silica reaction: a review. Cem. Concr. Res. **41**, 1224–1231 (2011)
240. Thomas, R.: Practical Guide to ICP-MS, 2nd Edn., p. 376. CRC Press, Boca Raton/FL (2008)
241. Tkachev, Y.A.: Study of mathematical statistics by geology students: necessary minimum (in Russian, with English abstract.). Institute of Geology of Komi Science Center of Ural Branch Russian Academy of Sciences, Vestnik, **239**, 24–29 (2014)
242. Tritthart, J.: Chloride binding in cement - I. Investigations to determine the composition of pore water in hardened cement. Cem. Concr. Res. **19**(4), 586–594 (1989)
243. Vallée, M.: Sampling quality control. Explor. Min. Geol. **7**, 107–116 (2000)
244. Van Aardt, J.H.P., Visser, S.: Formation of Hydrogarnets: calcium hydroxide attack on clays and feldspars. Cem Concr. Res. **7**(1), 39–44 (1977)
245. Van Aardt, J.H.P., Visser, S.: Calcium hydroxide attack on feldspars and clays: possible relevance to cement-aggregate reactions. Cem. Concr. Res. **7**(6), 643–648 (1977)
246. Van Aardt, J.H.P., Visser, S.: Reaction of $Ca(OH)_2$ and of $Ca(OH)_2 + Ca(SO_4) \cdot 2H_2O$ at various temperatures with feldspars in aggregates used for concrete mixing. Cem. Concr. Res. **8**(6), 677–682 (1978)
247. Van Gaans, P.F.M., Vriend, S.P., Schuiling, R.D.: Integral rock analysis: a new approach in lithogeochemical exploration with use of X-ray fluorescence spectrometry. Geol. Mijnbouw **65**, 205–213 (1986)
248. Van Grieken, R., Van De Velde, R., Robberecht, H.: Sample contamination from a commercial grinding unit. Anal. Chim. Acta **118**, 137–143 (1980)
249. Van Grieken, R., Markowicz, A.: Handbook of X-ray Spectrometry, 2nd Edn., p. 1016. CRC Press, Boca Raton/FL (2001)
250. Van Loon, J.C., Barefoot, R.R.: Analytical Methods for Geochemical Exploration, p. 344. Academic Press, San Diego/CA (1989)

251. Van Loon, J.C., Parissis, C.M.: Scheme of silicate analysis based on lithium metaborate fusion followed by atomic-absorption spectrophotometry. Analyst **94**(1125), 1057–1062 (1969)
252. Varma, A.: Handbook of Inductively Coupled Plasma Atomic Emission Spectroscopy, p. 380. CRC Press, Boca Raton/FL (1991)
253. vlab.amrita.edu: Flame photometry, To determine the concentration of alkali and alkaline earth metals in various samples (2012). https://vlab.amrita.edu/?sub=2&brch=193&sim=1351&cnt=1. Accessed 6 June 2024
254. Vollpracht, A., Lothenbach, B., Snellings, R., Haufe, J.: The pore solution of blended cements: a review. Mater. Struct. **49**(8), 3341–3367 (2016)
255. Von Ardenne, M.: Das Elektronen-Rastermikroskop. Theoretische Grundlagen, Zeitschrift für Physik **109**(9–10), 553–572 (1938)
256. Wang, E., Fengnian, S., Manlapig, E.: Mineral liberation by high voltage pulses and conventional comminution with same specific energy levels. Mineral. Eng. **27–28**, 28–36 (2012)
257. Wang, Y., Deng, M., Tang, M.: Alkali release from aggregate and the effect on AAR expansion. Mater. Struct. **41**, 159–171 (2008)
258. Wang, Y., Yu, G., Deng, M., Tang, M., Lu, D.: The use of thermodynamic analysis in assessing alkali contribution by alkaline minerals in concrete. Cem. Concr. Compos. **30**(4), 353–359 (2008)
259. Wang, X.T., Chen, Y., Zhou, Y.Q., Wang, M., Chen, X.L.: An effective method for making thin section of fluid inclusions in rock salt. Acta Mineral. Sinica **36**(3), 397–403 (2016)
260. Way, S.J., Cole, W.F.: Calcium hydroxide attack on rocks. Cem. Concr. Res. **12**(5), 611–617 (1982)
261. Weatherhead, A.V.: Petrographic Micro-Technique, p. 102. Arthur Barron Ltd., London/UK (1947)
262. Welz, B., Vale, M.G.R.: Atomic absorption spectrometry and related techniques. In: Cazes, J. (ed.) Ewing's Analytical Instrumentation Handbook, 3rd Edn., pp. 75–126. Marcel Dekker, New York/USA (2005)
263. Winkler, E.M.: Stone: Properties, Durability in Man's Environment, p. 230. Springer, Wien/AT (1973)
264. Williams, G.P.: Electron binding energies of the elements. In: Lide, D.R. (ed.) Handbook of Chemistry and Physics, Internet version 2005, pp. 10.200–10.206. CRC Press, Boca Raton/FL (2005)
265. Wong, H.S., Buenfeld, N.R.: Monte Carlo simulation of electron-solid interactions in cement-based materials. Cem. Concr. Res. **36**(6), 1076–1082 (2006)
266. Wong, H.S., Poole, A.B., Wells, B., Eden, M., Barnes, R., Ferrari, J., Fox, R., Yio, M.H.N., Copuroglu, O., Guðmundsson, G., Hardie, R.: Microscopy techniques for determining water–cement (w/c) ratio in hardened concrete: a round-robin assessment. Mater. Struct. **53**(25), 19 (2020)
267. Yanguas, J.E., Dravis, J.J.: Blue fluorescent dye technique for recognition of microporosity in sedimentary rocks. J. Sediment. Petrol. **55**(2), 600–602 (1985)
268. Zinovyev, N.T., Siomkin, B.V., Tanbayev, Z.G.: On concrete demolition by electric discharges. In: Kasai, Y. (ed.) Demolition and Reuse of Concrete and Masonry, pp. 300–305. Chapman & Hall, London/ UK (1988)
269. Zou, H.: Error propagation. In: McDonough, W.F. (ed.) Analytical Geochemistry/Inorganic Instrumental Analysis, 2nd Edn., Treatise on Geochemistry, **15**(2), pp. 33–42 (2014)

Chapter 4
International Standards and Recommendations

João Custódio and António Santos Silva

Abstract As noted in the previous chapters, concrete alkali content is a major factor determining expansion due to alkali-silica reaction (ASR). Limiting concrete alkali loading is imperative to reduce the risk of deleterious ASR development in structures, whenever potentially alkali-reactive or alkali-reactive aggregates are used in the concrete. Therefore, this chapter provides a brief overview of the current panorama on standards and recommendations regarding alkali content in concrete. If focuses on standards and recommendations that specify limits for concrete alkali loading and highlights the main standards and recommendations for determining the alkali content of concrete constituents and hardened concrete. The procedures for the analyses and, whenever available, alkali content limiting values for concrete constituents, are also presented.

Keywords Standards · Recommendations · Alkali loading · Alkali content determination methods · Concrete · Cement · Supplementary cementitious materials · Aggregates · Water · Additives · Chemical admixtures

---

This chapter was written with the cooperation of Bruno Godart, Eric Giannini, Esperanza Menéndez Méndez, Jason Ideker and Kazuo Yamada.

J. Custódio (✉) · A. Santos Silva
National Laboratory for Civil Engineering (LNEC), Av. do Brasil 101, 1700-066 Lisboa, Portugal
e-mail: jcustodio@lnec.pt

A. Santos Silva
e-mail: ssilva@lnec.pt

E. Menéndez Méndez and A. Santos Silva (eds.), *Total Alkalis in Concrete—Contribution to Alkali-silica Reaction*, RILEM State-of-the-Art Reports 40, https://doi.org/10.1007/978-3-031-56164-1_4

4.1 Standards and Recommendations for Limiting the Alkali Content in Concrete

As noted in the previous chapters, concrete alkali content is a major factor determining expansion due to alkali-silica reaction (ASR). In general, the higher the alkali level, the greater will be the expansion. Because of that, existing guides and recommendations, on the prevention of ASR in concrete structures, limit the concrete alkali loading whenever potentially alkali-reactive or alkali-reactive aggregates are used in the concrete. For the calculation of the concrete alkali loading, i.e., the total amount of alkalis in the concrete mixture expressed in kg/m^3, some guides consider only the alkalis deriving from the Portland cement [1, 23], some include the alkalis from the cement, type II additions and limestone aggregate [89], while others consider all concrete mix constituents, e.g., including also the chemical admixtures and mixing water [28, 59, 72, 82, 97, 98].

There are several standards and recommendations specifying limits for concrete alkali loading to reduce the risk of deleterious ASR in concrete. For instance, ASTM C1778 [23], AASHTO R 80 [1] and CSA A23.2-27A [40], in their prescriptive approach for selecting appropriate preventive measures define maximum permissible alkali loadings, according to the specific prevention level, with values as low as 1.2 or 1.8 kg/m^3 (Table 4.1); this prescriptive approach is not applicable when the alkali content of the Portland cement exceeds 1.25 wt%.

The Concrete Society Technical Report No. 30 [105] and the BRE Digest 330 [28] give advice for cements and combinations, including ground granulated blast-furnace slag (GGBS) and fly ash (FA), used with aggregates classified in three reactivity classes, for general concreting applications. It provides several limits to concrete alkali loading, depending on the specific concrete mixture (Tables 4.2 and 4.3). British Standards BS 8500-1:2015+A2:2019 [35] and BS 8500-2:2015+A2:2019 [36] also provide guidance on concrete alkali loading according to the aggregates'

Table 4.1 Maximum alkali loadings per level of prevention in North America [23, 40]

Level of prevention[c]	Maximum alkali loading of concrete (kg Na_2O_{eq}/m^3)
V	No limit
W	3.0
X	2.4
Y	1.8
Z[a]	1.8 (ASTM C1778), 1.2 (CSA A23.2-27A)
ZZ[a, b]	

Notes [a] SCMs may be used in prevention levels Z and ZZ

[b] SCMs cannot be used as sole prevention in level ZZ

[c] The details on levels of prevention and prescriptive approach are described in (ASTM C1778) and in (CSA A23.2-27A). This table shall not be used out of the complete context given in ASTM C1778 and in CSA A23.2-27A

reactivity (determined with the concrete prism tests specified in BS 812-123 [29] and use of supplementary cementitious materials (SCMs) (Table 4.4).

CEN Report CR 1901 [38] provides regional specifications and recommendations for the avoidance of damaging ASR in concrete including limits to concrete alkali loading (Table 4.5). French standard FD P18-464 [59] suggests limits to alkali loading of concretes without additions and using cements CEM I, CEM II, CEM III/A with less than 60 wt% slag, CEM V or concretes with additions having similar compositions (Table 4.6). RILEM [97, 98] and Portuguese LNEC Specification E 461 [82] provide recommendations on concrete alkali loading based on aggregate reactivity (Tables 4.7 and 4.8). In JIS A 5308 [72], for not-innocuous aggregate judged by chemical method [70] and/or mortar bar test [71], it is required to limit the total alkali content below 3.0 kg/m^3, or use blended cement (Type B)

Table 4.2 Recommended limits for the alkali content of concrete in UK [28, 105]

Aggregate type or combination classification#	Low alkali cement* ($\leq$0.6 wt% Na_2O_{eq})	Moderate alkali cement* ($\leq$0.75 wt% Na_2O_{eq})	High Alkali cement* (>0.75 wt% Na_2O_{eq})
Low reactivity	No limit†	No limit†	$\leq$5.0 kg Na_2O_{eq}/m^3‡
Normal reactivity	No limit†	$\leq$3.5 kg Na_2O_{eq}/m^3‡	$\leq$3.0 kg Na_2O_{eq}/m^3‡
High reactivity	$\leq$2.5 kg Na_2O_{eq}/m^3‡		

Notes * Alkali content of the CEM I-type component of the binder or the CEM I component of a combination with ground granulated blast furnace slag, pulverized fuel ash or fly ash

† The details on the self-limiting conditions and associated recommendation are described in [28, 105]

‡ The alkali contents presented are calculated in accordance with [28, 105]

This table shall not be used out of the complete context given in [28, 105]

Table 4.3 Recommended limits for the alkali content of concrete containing metakaolin and silica fume in UK [28]

Aggregate type or combination classification#	Low alkali cement ($\leq$0.6 wt% Na_2O_{eq})	Moderate alkali cement ($\leq$0.75 wt% Na_2O_{eq})	High Alkali cement (>0.75 wt% Na_2O_{eq})
Low reactivity	No limit†	No limit†	$\leq$5.0 kg Na_2O_{eq}/m^3‡, a, b
Normal reactivity	No limit†	$\leq$5.0 kg Na_2O_{eq}/m^3‡, a $\leq$4.0 kg Na_2O_{eq}/m^3‡, b	$\leq$5.0 kg Na_2O_{eq}/m^3‡, a $\leq$3.5 kg Na_2O_{eq}/m^3‡, b
High reactivity	$\leq$5.0 kg Na_2O_{eq}/m^3‡, a $\leq$3.0 kg Na_2O_{eq}/m^3‡, b		

Notes † The details on the self-limiting conditions and associated recommendation are described in [28]

‡ The details on alkali contents are given in [28]

This table shall not be used out of the complete context given in [28]

a With 10 wt% of metakaolin

b With $\geq$8 wt% of silica fume

Table 4.4 Recommended limits for the alkali content of concrete in UK [36]

Aggregate type or combination classification#	Maximum alkali loading of concrete (kg Na_2O_{eq}/m^3)
Low reactivity aggregate	≤5.0†
Normal reactivity aggregate	≤3.5† ≤4.0 (with ≥8 wt% of silica fume)†
High reactivity aggregates	≤2.5† ≤3.0 (with ≥8 wt% silica fume)†
Extreme reactivity aggregates	Not covered in the standard

Notes † The details on aggregate reactivity, concrete alkali contents and exposure classes are described in BS 8500-2 [36]. The alkali contents presented are calculated in accordance with BS 8500-2:2015+A2:2019 [36]
This table shall not be used out of the complete context given in BS 8500-2:2015+A2:2019 [36]

specified by JIS R 5211 [79] for slag cement and JIS R 5213 [80] for fly ash cement, respectively. The JIS standards for blended cements do not include regulations on alkali content because the only available fly ash in the Japanese market is siliceous with a reactivity index greater than 90 [73]. These Japanese fly ashes differ from high-CaO fly ashes (e.g., certain ASTM C618 Class C fly ashes) where the alkali content may increase the risk of ASR.

Further advice on ASR can be found in several reports and recommendations, albeit without direct recommendations for concrete alkali content, some examples are CSA A864-00 [41], FD P18-541 [60], FD P18-542 [61], ISO 22965-1 [65], ISO 22965-2 [66], CEN Report PD CEN/TR 16349 [93], CEN Report PD 6682-1 [92], IStructE Technical guidance on ASR [69], ACI 201.2R-16 [2].

The purpose of the following sections of this chapter is to outline the main standards and recommendations for determining the alkali content of concrete constituents and hardened concrete. They will briefly outline the procedure of the analysis and, whenever available, alkali content limiting values for concrete constituents will also be considered.

Although, all industrialized countries have their own national standards, defining the permitted composition for concrete constituents and test procedures to determine their composition and properties, many countries have been strongly influenced by those developed in Europe by the European Committee for Standardization (CEN)

Table 4.5 Recommended limits for the alkali content of concrete [38]

Category and class of cement#	Maximum alkali loading of concrete (kg Na_2O_{eq}/m^3)
CEM I	≤3.0†
CEM III A	≤4.5 (with <50 wt% of slag)† ≤5.5 (with >50 wt% of slag)†
CEM III B	≤10.0†
CEM III C	≤10.0†

Notes † The details on concrete alkali contents are described in [38]
This table shall not be used out of the complete context given in [38]

Table 4.6 Recommended limits for the alkali content of concrete in France [59]

Concrete	Maximum alkali loading of concrete (kg Na_2O_{eq}/m^3)
Concretes without additions and using cements CEM I, CEM II, CEM III/A with less than 60 wt% slag, CEM V or concretes with additions having similar compositions	$T_{avg} < \frac{3.5}{1+2\left[\frac{\sqrt{C^2\sigma_C^2+S^2\sigma_S^2+V^2\sigma_V^2}}{CA_{C,avg}+SA_{S,avg}+VA_{V,avg}}\right]}$ *and* $T_{max} < 3.5^{\dagger}$ *or* $T_{avg} < 3.0$ *and* $T_{max} < 3.3^{\dagger}$ *Where:* $T_{avg} = \frac{CA_{C,avg}+BA_{B,avg}+UA_{U,avg}+EA_{E,avg}+HA_{H,avg}}{100}\,\dagger$ $T_{max} = \frac{CA_{C,max}+BA_{B,avg}+UA_{U,avg}+EA_{E,avg}+HA_{H,max}}{100}\,\dagger$
Concretes using a blast furnace cement CEM III/A with more than 60 wt% slag or a CEM III/B	$a_{avg} < 1.1$ *wt%*† *(where* a_{avg} *represents the average total alkali content of the cement)*†
Concretes using a blast furnace cement CEM III/C	a_{avg} *(CEM III/C) < 2 wt%*† *(where* a_{avg} *represents the average total alkali content of the cement)*†

Notes † The details on concrete alkali contents are described in [59]. Abbreviations: T_{avg} - average active alkali content of concrete (kg/m^3); T_{max} - maximum active alkali content of concrete (kg/m^3); a_{avg} - average total alkali content for cements CEM III (wt%); B - quantity of aggregates (kg/m^3); C - quantity of cement (kg/m^3); E - quantity of water (kg/m^3); H - quantity of additions (kg/m^3); U - quantity of admixtures (kg/m^3); A_B - content of active alkalis of aggregates (kg/m^3); A_C - content of active alkalis of cement (wt%); A_E - content of active alkalis of water (wt%); A_H - content of active alkalis of additions (wt%); A_U - content of active alkalis of admixtures (wt%). This table shall not be used out of the complete context given in [59]

Table 4.7 Recommended limits for the alkali content of concrete [97]

Aggregate type or combination	Maximum alkali loading of concrete (kg Na_2O_{eq}/m^3)
Low reactivity aggregate	No requirement†
Medium reactivity aggregate	≤3.0–3.5†
High reactivity aggregates	≤2.5–3.0†

Notes † The details on aggregate reactivity, concrete alkali contents, safety factors for concrete alkali loading, and need to use type II additions are described in [97]. The alkali contents presented are calculated in accordance with [97]. This table shall not be used out of the complete context given in [97]

Table 4.8 Recommended limits for the alkali content of concrete in Portugal [82]

Aggregate type or combination	Maximum alkali loading of concrete (kg Na_2O_{eq}/m^3)
Reactivity class I	No requirement†
Reactivity class II	≤3.0†
Reactivity class III	≤2.5†

Notes † The details on aggregate reactivity, concrete alkali contents, safety factors for concrete alkali loading, and need to use type II additions are described in [82]. The alkali contents presented are calculated in accordance with [82]. This table shall not be used out of the complete context given in [82] and in North America by the Canadian Standards Association and by ASTM International. Because of that most of the existing national standards dealing with methods to quantify the total amount of alkalis in concrete constituents are either issued by those institutions or derive from them as National versions, some exceptions are the standards produced by the Association Française de Normalisation (AFNOR) and the Japanese Standards Association Group (JSA).

4.2 Standardized Method to Quantify the Total Amount of Alkalis

4.2.1 *Cements*

Cement, due to its typical chemical composition and dosage used in the concrete, is a major source of alkalis in the concrete for the development of the ASR, thus cement alkali level is a key factor determining expansion due to ASR.

Generally, if an aggregate is considered nonreactive by petrographic analysis or shown to be non-reactive or innocuous by producing little or no expansion in the accelerate mortar bar test (AMBT) or in the concrete prism test (CPT), no mitigation measures are necessary and, therefore, cements with moderate or high alkali content can be used in the concrete, from the standpoint of ASR. Similarly, if the aggregate has a long satisfactory service field record with similar cementitious materials having similar or higher alkali levels, no mitigation is usually deemed necessary. On the other hand, use of aggregates judged to be potentially deleteriously alkali silica-reactive should be considered with the use of measures known to prevent excessive expansion such as restricting the alkali content of the concrete. This can be achieved by the use of cement with a low effective alkali content, although in France this measure by itself is not recommended [59], or by the use of an adequate proportion of slag, fly ash, silica fume or other pozzolanic material (in cement or as an addition) [23, 28, 72, 82, 97, 105].

In spite of this, most of the existing standard specifications for cement simply state that a cement shall conform to the respective chemical requirements prescribed in the standard; however, these do not include a limit value for the cement alkali content, the exception being the JIS R 5210 [78]. The use of low-alkali cements has been found to be effective in some cases in preventing ASR damage. Though, this precautionary measure may not be effective for concretes with unusually high cement contents, for concretes with significant alkalis from constituents other than the cement, if there are significant sources of extraneous alkalis (e.g., de-icing chemicals), for highly reactive aggregates or if the passage of moisture concentrates the alkalis in certain parts of the concrete structure [23, 28, 82, 97, 105]. Because of that, currently, there is still no agreed value at European level for the allowable alkali content of the cement. A summary of current requirements and recommendations concerning ASR and binder in the concrete is given in Tables 4.9 and 4.10.

The approach followed in ASTM C595/C595M [12], which deals with blended hydraulic cements for both general and special applications (using slag, pozzolan, limestone, or some combination of these, with Portland cement or Portland cement clinker or slag with lime). This Standard Specifications refers to the user to follow ASTM C1778 [23] to avoid the deleterious development of ASR. It notes that the purchaser can request the alkali content of the Portland cement, slag, fly ash, natural pozzolan or silica fume in the blended cement so that ASTM C1778 can be followed. ASTM C1157/C1157M [17], a performance specification covering hydraulic cements for both general and special applications, provides an optional requirement for resistance to ASR, when the cement is to be used with alkali-reactive aggregate, namely a limit for the expansion observed at 14 days (0.020%) and at 56 days (0.060%) in the ASTM C227 test method [8]. It should be noted that ASTM C227 is no longer an active standard due to the many issues with the reliability of the method mainly owing to high amounts of leaching in the small cross section (25 mm × 25 mm) bar and the storage over water at 38 °C. ASTM is in the process of updating all standards to refer to ASTM C1778 and to remove information regarding ASR risk management from test methods and specifications so that users get a clear and consistent information.

EN 197-1 [44] does not specify any requirement for dealing with ASR, it merely states that adequate requirements are given in EN 206 [45] which, in turn, recommends that if aggregates contain varieties of silica susceptible to be attacked by alkalis and the concrete is exposed to humid conditions, actions shall be taken to prevent deleterious ASR using provisions valid in the place of use (e.g., LNEC Specification E 461 [82] in Portugal). The EN 206 standard also advises that precautions appropriate to the geological sources of the aggregates should be undertaken, long-term experience with the particular combination of cement and aggregate shall be taken into account, and informs that a framework for specifying requirements for minimizing the risk of damaging alkali-silica reaction can be found in the CEN technical report TR 16349 [93].

The procedure that should be followed to determine the cement alkali content, typically expressed as sodium oxide equivalent (Na_2O_{eq}), obtained by converting the cement potassium oxide content (K_2O) to the equivalent cement sodium oxide content and adding the measured sodium oxide content (Na_2O) using the formula percent $Na_2O + 0.658$ wt% K_2O, is described in several standards. The most widespread methods are based on the traditional analysis by wet chemistry and on instrumental chemical analysis by X-ray fluorescence, e.g., ASTM C114 [6], EN 196-2 [43], JIS R 5202 [76], JIS R 5204 [77], ISO 29581-1 [67], ISO 29581-2 [68]. The Na_2O_{eq}, determined in accordance with these standards, is sometimes referred to as the acid-soluble alkali content to differentiate it from the sodium oxide equivalent obtained with determinations carried out using other extraction methods, e.g., water-soluble alkali content (also covered in ASTM C114 [6]).

Normally, the cement alkali content considered to calculate the concrete alkali loading is the cement total alkali content (i.e. the acid soluble alkali content),

Table 4.9 Standard specifications for cements and ASR precautions

Cement type	Standard specification	ASR related requirements	Applicable test method
Portland	ASTM C150/C150M [7]	The equivalent alkali content (Na_2O_{eq}) is to be reported on the manufacturer's certification. An explanatory note is provided stating that cements with equivalent alkalis below 0.6% by weight had historically been designated as "low-alkali cements" but that users should now refer to ASTM C1778 for guidance on maximum alkali loading when there are concerns about reactive or potentially reactive aggregates.	ASTM C114 [6]
Blended hydraulic cements	ASTM C595/C595M [12]	Users are directed to ASTM C1778 for guidance on mitigating ASR with ASTM C595 blended cements.	ASTM C1778 [23]
Hydraulic cements	ASTM C1157/C1157M [17]	If the cement is to be used in concrete with alkali-silica reactive aggregate, the expansion obtained in the mortar-bar test should be less than 0.020% at 14 days and less than 0.060% at 56 days.	ASTM C227 [8]
Common cements	EN 197-1 [44]	Informs that alkalis from cement or other concrete constituents may react chemically with certain aggregates, and that adequate requirements are given in EN 206.	–
Very low heat special cements	EN 14216 [56]	–	–
Supersulfated cement	EN 15743:2010+A1:2015 [58]	–	–
CEM III/A with more than 60 wt% slag CEM III/B	FD P18-464 [59]	$a_{avg} < 1.1$ wt% (Where a_{avg} represents the average total alkali content of the cement).	NF EN 196-2 [87]

(continued)

Table 4.9 (continued)

Cement type	Standard specification	ASR related requirements	Applicable test method
CEM III/C	FD P18-464 [59]	$a_{avg} < 2$ wt% (Where a_{avg} represents the average total alkali content of the cement)	NF EN 196-2 [87]
–	BS 8500-1:2015+A2:2019 [35]	–	–
–	BS 8500-2:2015+A2:2019 [36]	–	–
–	ISO 22965-1 [65]	–	–
–	ISO 22965-2 [66]	Deleterious ASR shall be avoided by using procedures specified in the national annex or the standard.	–
Portland	JIS R 5210 [78]	The equivalent cement sodium oxide content should be less or equal to 0.75 wt% (0.6 wt% for low alkali cement).	

Notes Further details on the requirements are given in in the respective publications. The symbol "–" means that the standard has no information on the topic. This table shall not be used out of the complete context given in the respective publications

determined according to ASTM C114 [6], EN 196-2 [43], etc. However, for type CEM II and CEM V cements, the French standard NF P18-454 [89] considers an alternative method, which contemplates the alkali content obtained with the test methods described in EN 196-2 [43] and in LPC method no. 48 [104] with the insoluble residue, determined by LPC method no. 48, as follows:

$$A = 0.17(a_t - a_s) + \frac{(100 - J) + 0.5(J - r)}{100 - r} a_s$$

where: a_t is the equivalent cement sodium oxide content (wt%), determined according to EN 196-2 [43]; a_s is the equivalent cement sodium oxide content (wt%), determined according to LPC method no. 48; r is the insoluble residue content (wt%), determined according to LPC method no. 48; J is the sum of the cement constituents content other than the clinker.

Table 4.10 Guides, recommendations and reports on cement and ASR

Cement type	Document	ASR related requirements	Applicable test method
CEM I	CEN CR 1901 [38]	$Na_2O_{eq} \leq 0.6$ wt%	–
CEM III/A		$Na_2O_{eq} \leq 0.9$ wt%(slag < 50 wt%) $Na_2O_{eq} \leq 1.1$ wt% (slag ≥ 50 wt%)	–
CEM III/B CEM III/C		$Na_2O_{eq} \leq 2.0$ wt%	–
CEM I	PD CEN/TR 16349 [93]	$Na_2O_{eq} \leq 0.6$ wt%	
–	TR 30 [105]	No direct requirements are made concerning the cement alkali content. The approach followed consists in limiting the concrete alkali content.	–
–	IStructE [69]	–	–
CEM I	BRE Digest 330 [28]	Concrete made with cement having a Na_2O_{eq} (acid soluble) less than 0.6 wt% are unlikely to be at risk from damaging ASR even when using a normally reactive aggregate. When the cement alkali content is ≤ 0.6 wt%, the alkalis contributed from other sources should not exceed 0.60 kg Na_2O_{eq}/m^3. When the cement alkali content is ≤ 0.75 wt%, the alkalis contributed from other sources should not exceed 0.20 kg Na_2O_{eq}/m^3. If this value is exceeded, it can be accommodated by reducing the alkali limit by subtracting the total amount contributed.†	EN 197-1 [44]
–	LNEC E 461 [82]	No direct requirements are made concerning the cement alkali content. The approach followed consists in limiting the concrete alkali content.	–
–	RILEM AAR-7.1&7.3 [97, 98]	No direct requirements are made concerning the cement alkali content. The approach followed consists in limiting the concrete alkali content.	–

(continued)

Table 4.10 (continued)

Cement type	Document	ASR related requirements	Applicable test method
–	ASTM C1778 [23]	Alkali loading is defined in this standard as that contributed by the Portland cement in a concrete mixture. The prescriptive approach option to mitigating ASR risk includes the recommended maximum alkali loading for each level of prevention. The level of prevention is determined based on the aggregate reactivity, ASR risk based on element type and exposure, and structure class. Options are presented for mitigating ASR risk through limiting alkali loading, use of SCMs, or a combination these measures. The performance-based approach option to mitigating ASR risk relies upon testing mitigation measures following ASTM C1567 or ASTM C1293. This approach is not intended to evaluate the effect of alkali loading limits.	ASTM C114 [6]
–	AASHTO R 80 [1]	This standard presents a methodology that generally follows that described in ASTM C1778.	–
–	CSA A23.2-27A [40]	This standard presents a methodology that is close to that described in ASTM C1778.	CSA A3001 [42]
–	ACI 221.1R-98 [3]	No specific recommendations on cement alkali content or alkali loading are given. This report does note that limiting cement alkalis to $\leq$ 0.60 wt% Na_2O_{eq} is not a guarantee against deleterious ASR development with reactive aggregates. The importance of alkali loading, of the concrete rather than cement alkali content, is also stated.	–
–	ACI 201.2R-16 [2]	No specific recommendations on cement alkali content or alkali loading are given. This report does note that limiting cement alkalis to $\leq$ 0.60 wt% Na_2O_{eq} is not a guarantee against deleterious ASR development with reactive aggregates. The importance of alkali loading, of the concrete rather than cement alkali content, is also stated, and references are made to more recent versions of standards and specifications than are found in ACI 221.2R-98.	–

Notes † Further details on alkali contents are given in [28]. The symbol "–" means that the publication has no information on the topic. This table shall not be used out of the complete context given in the respective publications

4.2.2 Supplementary Cementitious Materials

Supplementary cementitious materials (SCMs) are well known to reduce, or to mitigate, the ASR effects in hardened concrete. These materials include fly ash (FA), ground granulated blast-furnace slag (GGBS), silica fume (SF), metakaolin (MK) or natural pozzolans (NPoz). The composition and pozzolanic reactivity of such materials is a crucial point in its dosage when reactive aggregates are used in concrete. As shown in Table 4.11 the chemical composition of SCMs varies within a certain range. With exception of GGBS, silica and alumina prevail. Alkalis can be present in considerable amounts (e.g., NPoz) and, in general, potassium prevails over sodium.

Table 4.12 compares the reactivity of different SCMs, expressed as consumption rates of calcium hydroxide per gram of pozzolan. Despite the variation of results obtained within the same material type, MK and SF present the highest pozzolanic reactivity, which is generally reflected in the dosage of each SCM when ASR inhibition is conducted. It should be noted that currently there are more methods for assessing the reactivity of SCMs (see conclusions of RILEM TC-267 Tests for reactivity of supplementary cementitious materials). However, the efficiency of SCMs in terms of ASR suppressing depends mainly on their efficiency to reducing the alkalinity and the portlandite content in the pore solution.

Table 4.11 Main element oxide contents of selected SCMs from the literature, in weight percent (wt%)

Oxide	Fly ash		Blast-furnace slag		Silica fume		Metakaolin		Natural pozzolana	
SiO_2	57.80	49.20	33.50	35.50	96.94	94.6	52.06	52.1	71.11	52.12
Al_2O_3	26.50	23.34	12.50	12.20	0.52	0.30	45.29	41.00	11.79	18.29
$Fe_2O_3{}^T$	7.00	14.72	0.17	1.01	0.14	0.10	0.60	4.32	2.57	5.81
CaO	3.30	3.81	39.30	41.40	0.58	0.50	0.05	0.07	2.07	4.94
MgO	0.70	1.03	11.70	8.30	0.00	0.40	0.00	0.19	0.15	1.20
SO_3	1.40	1.47	1.59	–	0.13	–	0.00	–	–	–
Na_2O	0.10	0.69	0.48	0.58	0.04	0.15	0.21	0.26	1.66	1.48
K_2O	1.40	1.78	0.45	0.64	0.42	0.56	0.16	0.63	1.33	5.06
LOI	1.70	2.55	0.80	–	1.47	3.29	0.51	0.60	9.50	11.10
Sum total	99.50	98.58	100.49	99.63	100.24	99.90	98.38	99.17	100.18	100.00
Na_2O_{eq}	1.02	1.86	0.78	1.00	0.32	0.52	0.32	0.67	2.54	4.81
Reference	1	2	3	4	5	6	7	8	9	9

1: Waller et al. [106]
2: Shafaatian et al. [100]
3: Gifford and Gillott [63]
4: Fernández-Jiménez and Puertas [62]
5: Shehata and Thomas [101]
6: Lane and Ozyildirum [81]
7: Ramlochan et al. [96]
8: Wild and Khatib [107]
9: Massazza [85]

Table 4.12 Pozzolanic activity of some SCMs

Pozzolan	FA		GGBS	SF		MK	
Reactivity (mg $Ca(OH)_2$/g pozzolan)	875	504	300	427	1089	1050	860
Reference	Sabir et al. [99]	Pontes et al. [94]	Andriolo and Sgaraboza [4]	Sabir et al. [99]	Quarcioni et al. [95]	Sabir et al. [99]	Pontes et al. [94]

Shehata and Thomas [101] have shown that for a certain content of FA replacement the expansion increases with the increase in the alkali content of fly ash. Although this, at present no consensus exists on what proportion of its content should to be included into the calculation of the alkalis in concrete mix.

The Concrete Society Technical Report no. 30 [105], RILEM AAR-7.1 recommendation [97] and BS 8500-2:2015+A2:2019 [36] give some information for the active alkali content of FA and GGBS that should be included in the calculation of the "reactive" alkalis in the concrete (Table 4.13). Other countries recommend different values, for example France indicates for siliceous FA, SF, MK and NPoz an active alkali content equal to 17 wt% of their total alkali content, and 50 wt% for GGBS, limestone and siliceous additions (in the case of MK, it is also possible to use the active alkali content determined according to the XP P 18-544) [59]. The alkali contents of siliceous FA, SF, GGBS, NPoz should be determined according

Table 4.13 Alkali contributions for FA and GGBS [36, 97, 105]

Addition type	Proportion (wt%) of addition in cement	Alkali contribution (wt%) to be included in alkali content of the concrete
FA (low lime with <8 wt% CaO and <5 wt% Na_2O_{eq})	>25	0
	20–25	20
	<20	100
GGBS (<1.5 wt% Na_2O_{eq})	≥40	0
	25–39	50
	<25	100

Note Where high reactivity aggregate is used and the cement or combination contains either GGBS or FA then proportions less than 50% by weight for GGBS and 40% by weight for FA are not recommended. The proportion of the alkali content of siliceous FA, where used as a filler aggregate, to be taken into account in the calculation of the alkali content of the concrete shall be determined in the same way as for FA used in cement or a combination. Further details on alkali contents are given in the respective publications. This table shall not be used out of the complete context given in the respective publications

Table 4.14 Standard specifications for the SCMs, recommendations and alkali limits

Addition type	Standard specification	Alkali limit (Na_2O_{eq})	Applicable test method
FA	EN 450-1 [46]	<5 wt%	EN 196-2 [43]
	ASTM C618 [13]	Should be declared but not specified†	ASTM C311/C311M [9]
SF	EN 13263-1:2005+A1:2019 [55]	Should be declared but not specified	EN 196-2 [43]
	ASTM C1240 [18]	Should be declared but not specified‡	ASTM C114 [6]
GGBS	EN 15167-1 [57]	Should be declared but not specified	EN 196-2 [43]
	ASTM C989/C989M [15]	Should be declared but not specified§	–
NPoz	ASTM C618 [13]	Should be declared but not specified†	ASTM C311/C311M [9]
MK	ASTM C618 [13]	Should be declared but not specified†	ASTM C311/C311M [9]
Blended	ASTM C1697 [22]	Should be declared but not specified*	ASTM C114 [6]/ASTM C311/C311M [9]
Ground glass	ASTM C1866/C1866M [25]	Type GS: ≤15.0 wt% Type GE: ≤4.0 wt%	ASTM C311/C311M [9]

Note † FA or NPoz meeting the requirement of expansion of test mixture as being equal or lower than that of low-alkali cement control mixture, at 14 days, are considered by the standard as effective in controlling ASR as the use of the low-alkali control cement used in the evaluation. However, it points out that the FA or NPoz shall be considered effective only when used at percentages by weight of the total cementitious material equal to or exceeding that used in the tests and when the alkali content of the cement to be used does not exceed that used in the tests by more than 0.05 wt% [13]. Furthermore, the standard mentions that the minimum amount of FA or NPoz should be 15% by weight of cementitious material [9]

‡ When the material is to be used with an aggregate that is regarded as being deleteriously alkali-reactive, the standard considers the option of determining the SF reactivity with alkalis, in accordance with ASTM test method C441, except that the amount of SF in the test mixture shall be 10% by weight of cementitious material. The reduction of mortar expansion resulting from the use of the SF shall be at minimum 80% [18]

§ If slag cement is to be used: (i) with a high-alkali Portland cement (Na_2O_{eq} > 0.60 wt%) or the concrete contains added water-soluble alkalis; and (b) with an aggregate that is regarded as deleteriously reactive with alkalis; then the standard considers necessary to test the effectiveness of slag cement in preventing excessive expansion due to ASR, using test method ASTM C441 [10]

* If required, blended supplementary cementitious materials shall conform to the optional requirements concerning their effectiveness in controlling ASR. The standard considers that materials meeting the expansion requirements are as effective in controlling alkali aggregate reactions as the use of the low-alkali control cement used in the evaluation. However, it is noted in the standard, that the materials shall be considered effective only when used at percentages by weight of the total cementitious material equal to or exceeding that used in the tests, and when the alkali content of the cement to be used does not exceed that used in the tests by more than 0.05%. The standard states that the materials shall be tested in accordance with Test Methods C311 or ASTM C441, depending on the blend type [22]. This table shall not be used out of the complete context given in the respective publications

the test-method defined in EN 196-2 [43], and shall be in conformity with the corresponding standard specifications (Table 4.14). As can be seen in Table 4.14, and with FA exception, the other SCMs have no indication of the values in alkalis that should not be exceeded in each specific standard.

According to BS 8500-2 [36], the alkali contents of GGBS, FA, SF, limestone fines and MK shall be determined in accordance with EN 196-2 [43] or by a secondary XRF calibrated against that method. The alkali content of siliceous FA used as a filler aggregate shall be determined by the appropriate method given in EN 196-2 or by a secondary XRF method calibrated against that method. The proportion of the alkali content of siliceous FA, if used as a filler aggregate, to be taken into account in the calculation of the alkali content of the concrete shall be determined in the same way as for FA used in cement or a combination.

A summary of current requirements and recommendations concerning SCMs alkali content and ASR is given in Tables 4.15 and 4.16. The values presented in Table 4.15 pertain to the prescriptive approach in ASTM C1778, AASHTO R 80 and CSA A23.2-27A, thus guidance on their use can be obtained in those documents.

According to JIS A 6201 [73], fly ash is classified in four types as shown in Table 4.17. There is no regulation on alkali content nor Ca content. Historically in Japan, only siliceous fly is assumed to be used. According to JIS A 6206 [75], ground granulated blast-furnace slag for concrete is classified in four types as shown in Table 4.18. There is no regulation on alkali content.

Table 4.15 Alkali content of SCM [1, 23, 40]

Type of SCM	Alkali content of SCM (wt%)	Minimum replacement level
Fly ash (CaO $\leq$18.0 wt%—ASTM C1778 and AASHTO R 80; CaO $\leq$ 15.0 wt%—CSA A23.2-27A)	<3.0	Varies according to level of prevention†
	3.0–4.0 (ASTM C1778) 3.0–4.5 (AASHTO R 80, CSA A23.2-27A)	
Slag cement	<1.0	
Silica fume (SiO_2 > 85 wt%)	<1.0	

Notes † Further details on alkali contents, replacement levels and levels of prevention are given in (ASTM C1778, AASHTO R 80, CSA A23.2-27A). This table shall not be used out of the complete context given in the respective publications

Table 4.16 Guides, recommendations and reports on alkali limits of SCMs

Document	Addition type	Alkali limit (Na_2O_{eq})	Applicable test method
BRE Digest 330 [28]	GGBS	<1.0 wt% Na_2O_{eq} Acid-soluble alkali content to be taken into account when calculating the contribution from the cement or combination: if GGBS proportion is <25 wt% then all alkalis are considered; if GGBS proportion is 25–39 wt% then all half of the alkalis are considered.	National Annex of BS EN 196-2
BRE Digest 330 [28]	FA	<5.0 wt% Na_2O_{eq} Total acid-soluble alkali content (determined following either a complete acid decomposition or a high temperature fusion) to be considered when calculating the contribution from the cement or combination: if FA proportion is <20 wt% then all alkalis are considered; if FA proportion is 20–24 wt% then a fifth of the alkalis are considered.	BS 3892-1
CEN CR 1901 [38]	–	–	–
PD CEN/TR 16349 [93]	–	–	–
CS Report TR 30 [105]	GGBS	Acid-soluble alkali content <1.0 wt% Na_2O_{eq}	National Annex of BS EN 196-2 or by a secondary XRF method calibrated against that method
	FA	Acid-soluble alkali content <5.0 wt% Na_2O_{eq}	BS 3892-1
IStructE [69]	–	–	–
LNEC E 461 [82]	–	–	–
RILEM AAR-7.1&7.3 [97, 98]	FA GGBS	Fly ash: <5.0 wt% Na_2O_{eq} Slag: <1.5 wt% Na_2O_{eq}	–
ASTM C1778 [23], AASHTO R 80 [1], CSA A23.2-27A [40]	FA, GGBS, SF	See respective guide for prescriptive or performance-based approach for managing the alkali content of the concrete and/or SCM.	–

Notes Further details on alkali contents and limits are given in the respective publications. This table shall not be used out of the complete context given in the respective publications. The symbol "–" means that the publication has no information on the topic

Table 4.17 Quality requirement for "Fly ash for use in concrete" JIS A 6201 [73]

	Type I	Type II	Type III	Type IV
SiO_2 (wt%)	≥45.0			
Moisture (wt%)	≤1.0			
LOI (wt%)	≤3.0	≤5.0	≤8.0	≤5.0
Density (g/cm)	≥1.95			
Fineness 45 μm over (wt%) Blaine surface area (cm^2/g)	≤10 ≥5000	≤40 ≥2500	≤40 ≥2500	≤70 ≥1500
Flow ratio (%)	105	95	85	75
Reactivity index (%) 28 days 91 days	90 100	80 90	80 90	60 70

Table 4.18 Quality requirement for "Ground granulated blast-furnace slag for concrete" JIS A 6206 [75]

	BBGS3000	GGBS4000	GGBS6000	GGBS8000
Density (g/cm)	≥2.80			
Specific surface area (cm^2/g)	2750 ≤ <3500	3500 ≤ <5000	5000 ≤ <7000	7000 ≤ <10000
Reactivity index (%) 7 days 28 days 91 days	– 60≤ 80≤	55≤ 75≤ 95≤	75≤ 95≤ –	95≤ 105≤ –
Flow ratio (%)	95≤	95≤	90≤	85≤
MgO (wt%)	≤10.0			
SO_3 (wt%)	≤10.0			
LOI (wt%)	≤3.0			
Cl (wt%)	≤0.02			
Basicity (CaO + MgO + Al_2O_3)/SiO_2	≥1.60			

4.2.3 Aggregates

In the last years, aggregates are being questioned as being an important source for the increase of alkalis content into concrete pore solution. Many aggregate types are very rich in alkali minerals, and depending of their composition and alteration degree, the extent to which these are released and the conditions for it remains uncertain. This subject has been studied by different authors [27, 39, 84] and methods of analysis were suggested in order to determine the content of alkalis that can be released when the aggregates are reacting with strong alkaline solutions, such as sodium and potassium hydroxide [26, 102, 103].

In scope of TC 258-AAA a new test method was proposed to assess the potential amount of alkalis released in the long term in concrete by aggregates through the determination of the amounts of sodium and potassium ions released by aggregates immersed in KOH and NaOH solutions, respectively, when in contact with excess calcium hydroxide [86].

Despite this evidence, the current European and American standard specifications for concrete aggregates [5, 52] do not present any limit values for the total alkali content or for the soluble alkali content (Table 4.19). Furthermore, no standard test methods for determining the aggregates' alkali content were found in the literature.

Although BRE Digest 330 [28] does not provide a limit for the aggregate alkali content, it adopts a conservative approach to the alkali contributions other than the binder (e.g., aggregate, admixture, non-potable mixing water), by taking all the alkali into account where the contribution exceeds 0.2 kg Na_2O_{eq}/m^3 and by limiting the maximum contribution to 0.60 kg Na_2O_{eq}/m^3. According to BRE Digest 330 [28], Concrete Society Technical Report no. 30 [105], JIS A 5308 [72] and the British standard BS 8500-2 [36], the alkalis contributed by salts or seawater in aggregates are calculated from the measured chloride ion concentration in the aggregate and

Table 4.19 Standard specifications for aggregate and ASR considerations

Standard specification	ASR related recommendations	Applicable test method
ASTM C33/C33M [5]	Users are directed to ASTM C1778 for guidance on assessing and mitigating ASR risk.	ASTM C1778 [23]
EN 12620:2002+A1:2008 [52]	The standard states that when required the alkali-silica reactivity of aggregates shall be assessed in accordance with the provisions valid in the place of use and the results declared. The standard also indicates that further information may be obtained in the CEN CR 1901 [38].	–
NF P 18-545 [90]	In the presence of aggregates or mixtures of potentially reactive aggregates, this standard recommends applying one of two solutions: (1) the balance of the active alkalis of all the constituents of the concrete; (2) a performance-based solution which comprises an expansion test (NF P 18-454 and FD P 18-456), to validate the concrete formula that will be implemented. This document also allows the use of reactive aggregates subject to applying the preventive provisions proposed in it. It allows the use of reactive aggregates in the category III works (e.g., big dams, tunnels, exceptional bridges or viaducts), whatever their class of exposure, since the supply of other type of aggregates is particularly difficult and that the reactive aggregates envisaged satisfy a preliminary formulation study.	XP P 18-544 [108]

Note This table shall not be used out of the complete context given in the respective publications. The symbol "–" means that the standard has no information on the topic expressed as equivalent sodium oxide: $wt\%Na_2O_{eq} = 0.76 \times wt\%Cl^-$ (where Cl^- is expressed as a percentage by weight of the dry aggregate, and 0.76 is a factor obtained from consideration of the composition of seawater). This applies to aggregates other than filler aggregate made from fly ash and coarse crushed concrete aggregate. These standards do not cover coarse recycled aggregate, fine recycled aggregate and fine crushed concrete aggregate. The chloride content of the aggregates shall be measured by the method given in EN 1744-1:2009+A1:2012 [51].

4.2.4 Water

Since the quality of the mixing water for production of concrete can influence setting time, strength development of concrete, protection of the reinforcement against corrosion, and the durability of the actual concrete, several standard specifications have been developed to provide compositional and performance requirements for water used as mixing water in hydraulic cement concrete (Table 4.20).

Typically, water supplied as potable water is deemed to conform to the requirements set in the applicable standard specifications, e.g., ASTM C1602/C1602M [20], EN 1008 [50], and therefore standard potable water is permitted to be used as mixing water in concrete without testing.

According to ASTM C1602/C1602M mixing water that is wholly or partially composed of sources of water that are non-potable or from concrete production operations is permitted to be used in any proportions to the limits qualified to meet the concrete performance requirements set in the standard specification. For sources of non-potable mixing water proposed for use as total mixing water or in the combined mixing water, the total combined mixing water must fulfil both the concrete performance requirements and the chemical limits, which include a limit of 600 ppm (0.06

Table 4.20 Standard specifications for mixing water and ASR considerations

Standard specification	ASR related requirements	Applicable test method
ASTM C1602/ C1602M [20]	The standard does not provide any recommendation on ASR. Nevertheless, it states that mixing water that is wholly or partially composed of sources of water that are non-potable or from concrete production operations must have an equivalent water sodium oxide content lower than 600 ppm.	ASTM C114 [6]
EN 1008 [50]	The standard states that the equivalent sodium oxide content of the water shall not exceed 1500 mg/L, unless actions have been taken to prevent deleterious alkali-silica reaction.†	EN 196-2 [43]

Note † CEN/TR 1901 gives a framework for specifying requirements for minimising the risk of damaging alkali-silica reaction. This table shall not be used out of the complete context given in the respective publications wt%) for the equivalent water sodium oxide content. This standard does not mention any particular requirement for water when is to be used in concrete with aggregates that are potentially alkali reactive. The equivalent water sodium oxide content should be determined according to the test method defined in ASTM C114 [6].

EN 1008 [50] has a similar approach to that of ASTM C1602/C1602M [20]; however, it adds that if alkali-reactive aggregates are expected to be used in the concrete, the water shall be tested for its alkali content in accordance with EN 196-2 [43]. EN 1008 [50] indicates that the equivalent sodium oxide content of the water shall not exceed 1500 mg/L (0.15%), and that if this limit is exceeded, the water may be used only if it can be shown that actions have been taken to prevent deleterious alkali-silica reactions. BS 8500-2:2015+A2:2019 [36] requires that the alkali content of non-potable water shall be determined by the methods given in BS 6068-2.42 [32] for sodium and BS 6068-2.43 [33] for potassium or by the method given in BS 6068-2.44 [34] for both sodium and potassium.

Concerning the use of sea water as mixing water in concrete, ASTM C1602/C1602M [20] does not address it directly, but it is known that sea water has an equivalent water sodium oxide content above 600 ppm. EN 1008 [50] allows the use of seawater as mixing water for concrete without reinforcement or other embedded metal, but considers it not being suitable for the production of reinforced or prestressed concrete. Concerning ASR, the 1500 mg/L limit for the equivalent sodium oxide content of the water indirectly prohibits seawater of being used whenever alkali-reactive aggregates are expected to be used in the concrete. Concrete Society Technical Report no. 30 [105] recommends that when the mixing water is non-potable or seawater, the alkali content should be determined in accordance with BS 6068 and expressed as the sodium oxide equivalent (Na_2O_{eq}) and included in the summation of alkali from other sources.

4.2.5 Additives

The alkalis coming from additives, i.e., finely-divided-inorganic constituents used in concrete in order to improve certain properties or to achieve special properties, like type I additions (nearly inert additions), should also be considered when dealing with ASR. The general suitability of additions for use in concrete is established for filler aggregate conforming to EN 12620:2002+A1:2008 [52], EN 13055-1:2002 [54] or ASTM C1797 [24]; and for pigments conforming to EN 12878 [53] or ASTM C979/C979M [14].

EN 12620 and EN 13055 do not specify a sodium oxide equivalent content limit, stating only that when required the alkali-silica reactivity of aggregates shall be assessed in accordance with the provisions valid in the place of use and the results declared. Furthermore, it is reported that in the absence of previous long-term experience of a lack of disruptive reactivity of a particular combination of cement and aggregate, it can be necessary to take one of the following precautions: (i) limit the total alkali content of the concrete mix; (ii) use a cement with a low effective alkali content; (iii) use a non-reactive aggregate combination; (iv) limit the degree of saturation of the concrete with water. The combination of aggregates and cement can be assessed using regulations applying at the place of use when compliance with one of the above procedures is not possible. According to BS 8500-2:2015+A2:2019 [36], where limestone fines are used as a filler aggregate and the determined alkali content is less than or equal to 0.1 wt% Na_2O_{eq}, no account shall be taken of its contribution to the alkali content of the concrete. The proportion of the alkali content of siliceous fly ash, where used as a filler aggregate, to be considered in the calculation of the alkali content of the concrete shall be determined in the same way as for fly ash used in cement or a combination. The alkali content of limestone fines and siliceous fly ash used as a filler aggregates shall be determined by in accordance with EN 196-2 [43] or by a secondary X-ray fluorescence method calibrated against that method. Although, the effect of pigments on ASR is not known, it is necessary to take into account the alkali content of any pigment used in concrete containing potentially alkali-reactive aggregate [105].

As mentioned before in Sect. 4.2.2, in France [59] the proportion of active alkalis with respect to the total alkalis is different for each of the constituents of the concrete. For siliceous additions, the active alkali content is the value obtained with the test method defined in XP P 18-544, or it is equal to 0.5 times the total alkali content determined according to NF EN 196-2 [87]. For limestone additions of category A, the active alkali content is equal to 0.5 times the total alkali content determined according to NF EN 196-2; for other limestone additions the active alkali content is equal to the total alkali content determined according to NF EN 196-2.

ASTM C979/C979M [14] and EN 12878 [53] make no recommendations towards ASR prevention. ASTM C1797 [24] makes no mention of ASR or prevention of ASR. Furthermore, no standard test methods to determine the alkali content of pigments were found in the literature.

4.2.6 Chemical Admixtures

Existing standards and specifications for chemical admixtures (Table 4.21), constituents to be added to hydraulic-cement concrete mixtures during the mixing process in small quantities related to the weight of cement to modify the properties of fresh or hardened concrete, do not specify any requirements concerning alkali-silica reaction [8, 11, 16, 19, 21, 48, 49]. In Japan, according to JIS A 6204 [74], the total alkali content supplied from chemical admixture in concrete should be below 0.3 kg/m^3. However, it is known that admixtures may contain alkalis in their composition, sometimes in relevant amounts, therefore whenever alkali-silica reaction may be an issue, admixtures that contain large amount of alkalis should be avoided and their alkali content should be determined and included in the concrete alkali loading [105]. Standards EN 934-2:2009+A1:2012 [49] and BS 8500-2:2015+A2:2019 [36] require that the alkali content of admixtures be determined by the method given in EN 480-12 [47] or be declared as the specified maximum value given in the constituent

Table 4.21 Standard specifications for admixtures and ASR considerations

Standard specification	ASR related requirements	Applicable test method
ASTM C1017/C1017M [16]	This standard informs that admixtures that contain relatively large amounts of alkali may contribute to reaction with some aggregates. It also states that complying with the requirements of the standard does not assure acceptability when used with alkali-reactive aggregates and some cements.	–
EN 934-2:2009+A1:2012 [49]	Alkali content should be declared.	EN 480-12 [47]
JIS A 6204 [74]	The equivalent sodium oxide content of chemical admixtures in concrete should be less or equal to 0.3 kg/m^3.	–

Note This table shall not be used out of the complete context given in the respective publications. The symbol "–" means that the standard has no information on the topic

material standard or the manufacturer's declared value. FD P 18-464 Standard [59] refers that the active alkali content of the admixture is equal to the total alkali content determined according to NF EN 480-12 [88].

4.2.7 Concrete

The alkali content of concrete should be calculated from the mix proportions and the determined alkali contents of each of the constituents, except in the case of ground granulated blast-furnace slag and fly ash constituents, where Table 4.13—also applies [36, 97, 105].

When dealing with existing concrete structures or if the alkali content of each of the constituents has not been determined, it may be of interest to determine the alkali content of the hardened concrete. In the case of concretes made with CEM I cements, or concretes containing ground granulated blast-furnace slag, under specific conditions, the British standard BS 1881-124 [30] describes the sampling procedure, treatment of sample, and analytical method to be used on a sample of concrete to determine its alkali content. Besides the method detailed in BS 1881-124, there are at least two other methods available to determine the concrete alkali loading, namely the ones described in GranDuBé [64] and in LNEC MI 115 [83].

The method described in BS 1881-124 [30] considers either 2 g of a separated test sample (the test sample is prepared by breaking the concrete sample and removing fine material by sieving through a 125 μm sieve until a sufficient quantity of fines has been produced) or 5 g of an analytical test sample (details given in the standard). The test sample is weighted into a 400 mL beaker, dispersed with 150 mL of distilled or deionized water, and 20 mL of concentrated nitric acid are added. Then the solution is boiled gently for 10 min, and allowed to cool. The solution is quantitatively transferred to a 500 mL volumetric flask. The solution is diluted to the mark with water, mixed thoroughly and filtered off through a dry medium ashless filter paper. Afterwards, 25 mL of the filtrate are pipetted into a 100 mL volumetric flask and 10 mL of aluminium solution are added. The solution is diluted to the mark with cold water and mixed thoroughly. The Na_2O and K_2O contents of this solution are determined by flame photometry.

The method defined in LNEC MI 115 [83] considers 5 g of an analytical test sample (prepared to pass a 300 μm sieve, further details are given in the publication). The alkali content is determined according to EN 196-2 [43].

The method provided in GranDuBé [64] considers 1 g of an analytical test sample (prepared to pass a 315 μm sieve, further details are given in the publication). The test sample is weighted into a 250 mL beaker, dispersed with 100 mL of water and a magnetic stirrer for 2 min. Then, 40 mL of concentrated nitric acid are added. The solution agitated again with a magnetic stirrer but for 30 min, at ambient temperature. The solution is filtered off through a dry medium ashless filter paper, and washed with water. The solution is collected to a volumetric flask and diluted to the 250 mL mark with cold water and mixed thoroughly. The Na_2O and K_2O contents of this solution are determined by flame photometry, atomic absorption spectrometry or inductively coupled plasma atomic emission spectroscopy. Instead of the acid attack, the method also considers an alternative attack by water. The methodology is identical, except that the test sample is weighted into a 250 mL beaker, dispersed with 200 mL of water and agitated with a magnetic stirrer for 24 h.

4.3 Standardized Methods to Determine the Releasable Alkalis

4.3.1 Aggregates

Besides the RILEM AAR-8 test-method [86], only two standardized methods to determine the releasable alkalis from aggregates were found, namely the Portuguese standard NP 1382 [91] and the French standard XP P18-544 [108].

The RILEM AAR-8 test-method, developed in the scope of the RILEM TC 258-AAA, is performed using an aggregate with particles between 4 mm and 0.125 mm, which were washed and dried at 105 ºC before the test. The aggregate is then immersed in two separate containers containing 0.7M KOH and 0.7M NaOH solutions, respectively, with excess calcium hydroxide, and placed in an oven at 38 ºC, using a solution-to-aggregate ratio of 4 L/kg. Periodically the test containers are agitated, and samples removed by pipetting for sodium and potassium determinations using the atomic absorption spectrometry (AAS) or inductively coupled plasma spectrometer (ICP-OES).

The method described in NP 1382 [91] considers an aggregate sample of 10 g (prepared to pass a 300 μm sieve). The sample is weighted into a 250 mL Erlenmeyer flask and 100 mL of water are added. Then the solution is heated, and gentle ebullition is maintained for 5 min. After that period, the flask is transferred to a water bath at room temperature for 1 h, with regular shaking. The solution is filtered through a

Büchner funnel which contains a well seated retentive, dry filter paper, into a 500 mL filtering flask, using a weak vacuum. After complete cooling the solution is diluted to the 500 mL mark. The Na_2O and K_2O contents of this solution are determined by flame photometry.

The method described in XP P18-544 [108] considers a weight of aggregates or siliceous fillers of 500 g. The test samples, neither crushed nor ground, and are placed in a stainless steel or polypropylene container and brought into contact with a saturated solution of boiling lime. After 7 h (±30 min) of contact, the aqueous suspension is removed, filtered and the filtrate is subjected after acidification and dilution to analysis by atomic absorption spectrometry, to measure sodium and potassium contents.

4.3.2 Supplementary Cementitious Materials

No standardized methods to determine the releasable alkalis of SCMs were found. The exception being ASTM C311/C311M [9] which has a procedure for calculating available alkali. It involves mixing a sample of the material with hydrated lime and water. It is stored at 38 ± 2 °C for 28 days. Then the sample is redissolved/exposed to an excess of water, acidified to keep the ions in solution and measured using flame photometry. The result is reported as equivalent percentage of sodium oxide (Na_2O_{eq}).

4.3.3 Others

No additional standardized methods to determine the releasable alkalis of concrete constituents were found.

4.4 Standardized Methods to Quantify the Total Alkalis in the Concrete

In France, the rules to prevent alkali-silica reaction are described in the standard FD P 18-464 [59]. One of the principles of these recommendations is to set a limitation of the alkalis content on concrete and not on cement. One of the reasons is that it isn´t necessary to fix a limit of the alkalis content on cement when these are using non-reactive aggregates. In addition, in some cases of use of very reactive aggregates, the use of low-alkali cements is not a sufficient guarantee to avoid the reaction. The ASR is clearly a question of “good or bad couple” between a given aggregate and a given cement.

In the standard FD P 18-464 the level of prevention B (applied for most of the buildings and civil engineering structures), propose as precautions the limitation of the quantity of alkalis in concrete. Article 6.3.2 of this standard describes the rules to be applied for this limitation.

All alkalis are not involved in the alkali aggregate reaction, because some of them remain trapped in crystal lattices. This is why the estimation of the alkalis considers only the active alkalis, that is to say those capable of passing into solution. The proportion of active alkalis with respect to the total alkalis is different for each of the constituents of the concrete.

4.4.1 Calculation of Active Alkali Content in Concrete

The active alkali content of the concrete T is equal to the sum of the active alkalis supplied by each of the constituents:

$$T = \frac{C\ A_{\mathrm{C}} + B\ A_{\mathrm{B}} + U\ A_{\mathrm{U}} + E\ A_{\mathrm{E}} + H\ A_{\mathrm{H}}}{100}$$

where:

B, quantity of aggregates, in kg/(m^3 of concrete)

C, quantity of cement, in kg/(m^3 of concrete)

E, quantity of water, in kg/(m^3 of concrete)

H, quantity of additions, in kg/(m^3 of concrete)

U, quantity of admixtures, in kg/(m^3 of concrete)

A_B, content of active alkalis of aggregates, in % by weight

A_C, content of active alkalis of cement, in % by weight

A_E, content of active alkalis of water, in % by weight

A_H, content of active alkalis of additions, in % by weight

A_H, content of active alkalis of admixtures, in % by weight.

4.4.2 Rule to Determine the Active Alkaline Content in Each Concrete Component

The rules for determining the active alkaline contents of each concrete component are given in Table 4.22.

Table 4.22 Methods for determining the active alkali content of the concrete constituents

Constituent	Method to be applied
Cement	Active alkali content according to Annex A of standard NF P 18-454 Total alkali content according to NF EN 196-2
Aggregate	Content of active alkalis determined according to XP P 18-544
Admixture	Active alkali content equal to the total alkali content determined according to NF EN 480-12
Water	Active alkali content equal to the total alkali content which is determined according to NF EN 196-2, in application of NF EN 1008
Slag	Active alkali content equal to 0.50 times the total alkali content determined according to NF EN 196-2
Siliceous fly ash	Active alkali content equal to 0.17 times the total alkali content determined according to NF EN 196-2
Silica fume	Active alkali content equal to 0.17 times the total alkali content determined according to NF EN 196-2
Siliceous Additions	Active alkali content determined according to XP P 18-544, or, if not considered, equal to 0.5 times the total alkali content determined according to NF EN 196-2
Limestone additions	For additions of category A, the active alkali content shall be 0.50 times the total alkali content determined according to NF EN 196-2 In other cases, the active alkali content is equal to the total alkali content determined according to NF EN 196-2
Metakaolin	Active alkali content determined according to XP P 18-544, or failing this, taken equal to 0.17 times the total alkali content determined according to NF EN 196-2
Pozzolan	Active alkali content equal to 0.17 times the total alkali content determined according to NF EN 196-2

4.4.3 Calculation of the Mean and Maximum Value of Active Alkali Content of Each Concrete Component

The mean and maximum values of the active alkali contents of each constituent are calculated from the product data sheets of each of the constituents. If these data are not available, these values are based on the measurements of the active alkali content obtained on a representative number of samples which is to be determined as a function of the nature of the constituents.

For cements, fly ashes, and slags, the mean and maximum values of the active alkaline contents are evaluated. For the other constituents, only the mean value is evaluated. The following formulas are applied to calculate the mean and maximum active alkali content of the concrete:

$$T_{\text{mean}} = \frac{CA_{\text{C,mean}} + BA_{B,\text{mean}} + UA_{U,\text{mean}} + EA_{E,\text{mean}} + HA_{H,\text{mean}}}{100}$$

$$T_{\max} = \frac{CA_{\text{C,max}} + BA_{B,\text{mean}} + UA_{U,\text{mean}} + EA_{E,\text{mean}} + HA_{H,\max}}{100}$$

$A_{H,max}$ being equal to the maximum active alkali content in the case of fly ash and slag and to the mean active alkali content in the case of other types of additions.

To evaluate these mean and maximum values, the Table 4.23 provides the rules for the statistical data to be provided for each of the constituents.

Table 4.23 Rules relating to the data to be provided to assess the alkali content of concrete constituents

Constituent	Evaluated content	Rules
Cement	$A_{C,mean}$ and $A_{C,max}$	Statistics over the last 12 months (with at least one value per month). In the absence of these values, ten samples distributed uniformly over a period of 3 months of production.
Aggregate	$A_{B,mean}$	Annual value. In the absence of this value, two samples should be taken from the lot in question.
Admixture	$A_{U,mean}$	Average of the 2 values obtained during the year under the "NF-Admixture" mark. In the absence of this value, the maximum of the content set in standard 934-2.
Recycling or catchment water	$A_{E,mean}$	Annual value. In the absence of this value, three samples should be taken over a month of production. NOTE: The supply of alkalis by the water in the drinking water system is considered as zero. Depending on the source, this could fluctuate greatly. Areas catching runoff from pavements, bridges, etc. with de-icing/anti-icing applications especially. May want to recommend sampling late winter/early spring in these situations to find maximum in these situations.
Fly ash and slag	$A_{H,mean}$ and $A_{H,max}$	Statistics over one year (with at least one value every two months). In the absence of these values, six samples should be taken over 3 months of production.
Limestone additions Class A	$A_{H,mean}$	Annual value. In the absence of this value, two samples should be taken from the lot concerned.
Other additions	$A_{H,mean}$	Annual value. In the absence of this value, two samples should be taken over 3 months of production.

4.4.4 Limits of the Alkali Content of Concrete

The slag, although itself containing alkalis, being very effective in reducing the risk of deleterious ASR, it has been found necessary to distinguish two cases, depending on whether the cements contain a high proportion of slag or not.

- *Concrete without additions, based on CEM I, CEM II, CEM III/A with less than 60 wt% slag, CEM V or concretes with additions having similar compositions (general case)*:

  (a) When the statistical data for the active alkaline contents of cement, slags and fly ash are available, the following two rules shall be verified:

$$T_{\text{mean}} < \frac{3.5}{1 + 2\left(\frac{\sqrt{C^2\sigma_C{}^2 + S^2\sigma_S{}^2 + V^2\sigma_V{}^2}}{CA_{C,mean} + SA_{S,\text{mean}} + VA_{V,mean}}\right)}$$

where:

C, quantity of cement, in kg/(m^3 of concrete)

S, quantity of slag, in kg/(m^3 of concrete)

V, quantity of fly ash, in kg/(m^3 of concrete)

σ_C, standard deviation of the active alkali content of cement, in % by weight

σ_S, standard deviation of the active alkali content of the slag, in % by weight

σ_V, standard deviation of the active alkali content of siliceous fly ash, in % by mas

and

$$T_{\text{max}} < 3.5\,\text{kg/m}^3$$

(b) In the absence of data on the dispersions of alkali contents, it is assumed that a concrete formula is suitable for the use of potentially reactive aggregates if the mean content T_{mean} is less than 3 kg per m^3 and the Maximum content T_{max} remains less than 3.3 kg per m^3.

- *Concrete without additions, based on CEM III/A cements with more than 60 wt% slag, CEM III/B and CEM III/C (particular case).*

  In this case, the total alkali content of these cements should be limited:

(a) Case of concretes made of CEM III/A with more than 60 wt% slag or CEM III/B: A concrete formula is suitable for the use of potentially reactive aggregates if: a_{mean} (CEM III/A or CEM III/B) <1.1 wt%.

Where a_{mean} (CEM III/A or CEM III/B) represents the average total alkali content of the cement used expressed as % of the cement weight.

(b) Case of concretes made of CEM III/C: A concrete formula is suitable for the use of potentially reactive aggregates if: a_{mean} (CEM III/C) <2 wt%.

Where a_{mean} (CEM III/C) represents the average total alkali content of the CEM III/C cement expressed as a % of the cement weight.

Acknowledgements The author João Custódio carried out this work within the scope of the FCT Investigator Project IF/00595/2015/CP1326/CT0001, financed by FCT - Fundação para a Ciência e a Tecnologia, Portugal. João Custódio wishes to acknowledge this support.

References

1. AASHTO R 80: Standard Practice for Determining the Reactivity of Concrete Aggregates and Selecting Appropriate Measures for Preventing Deleterious Expansion in New Concrete Construction, 25 pp. AASHTO, USA (2017)
2. ACI 201.2R-16: Guide to Durable Concrete, 84 pp. ACI, USA (2016)
3. ACI 221.1R-98: State of the Art Report on Alkali-Aggregate Reactivity, 31 pp. ACI, USA (1998)
4. Andriolo, F.R., Sgaraboza, B.C.: The use of pozzolan from calcined clays in preventing excessive expansion due to alkali-aggregate reaction in some brazilian dams. In: Proceedings of the 7th International Conference on Alkali Aggregate Reaction, 1986, Ottawa, Canada, pp. 66–70 (1987)
5. ASTM C33/33M: Standard Specification for Concrete Aggregates, 8 pp. ASTM, USA (2018)
6. ASTM C114: Standard Test Methods for Chemical Analysis of Hydraulic Cement, 33 pp. ASTM, USA (2018)
7. ASTM C150/C150M: Standard Specification for Portland Cement, 9 pp. ASTM, USA (2022)
8. ASTM C260/C260M: Standard Specification for Air-Entraining Admixtures for Concrete, 4 pp. ASTM, USA (2010)
9. ASTM C311/C311M: Standard Test Methods for Sampling and Testing Fly Ash or Natural Pozzolans for Use in Portland-Cement Concrete, 11 pp. ASTM, USA (2022)
10. ASTM C441/C441M: Standard Test Method for Effectiveness of Pozzolans or Ground Blast-Furnace Slag in Preventing Excessive Expansion of Concrete Due to the Alkali-Silica Reaction, 5 pp. ASTM, USA (2018)
11. ASTM C494/C494M: Specification for Chemical Admixtures for Concrete, 10 pp. ASTM, USA (2020)
12. ASTM C595/C595M: Standard Specification for Blended Hydraulic Cements, 10 pp. ASTM, USA (2022)
13. ASTM C618: Standard Specification for Coal Fly Ash and Raw or Calcined Natural Pozzolan for Use in Concrete, 5 pp. ASTM, USA (2022)
14. ASTM C979/C979M: Standard Specification for Pigments for Integrally Coloured Concrete, 5 pp. ASTM, USA (2016)
15. ASTM C989/C989M: Standard Specification for Slag Cement for Use in Concrete and Mortars, 7 pp. ASTM, USA (2022)
16. ASTM C1017/C1017M: Standard Specification for Chemical Admixtures for Use in Producing Flowing Concrete, 9 p. ASTM, USA (2022)
17. ASTM C1157/C1157M: Standard Performance Specification for Hydraulic Cement, 5 pp. ASTM, USA (2020)
18. ASTM C1240: Standard Specification for Silica Fume Used in Cementitious Mixtures, 7 pp. ASTM, USA (2020)

19. ASTM C1582/C1582M: Standard Specification for Admixtures to Inhibit Chloride-Induced Corrosion of Reinforcing Steel in Concrete, 10 pp. ASTM, USA (2017)
20. ASTM C1602/C1602M: Standard Specification for Mixing Water Used in the Production of Hydraulic Cement Concrete, 5 pp. ASTM, USA (2022)
21. ASTM C1622/C1622M: Standard Specification for Cold-Weather Admixture Systems, 7 pp. ASTM, USA (2017)
22. ASTM C1697: Standard Specification for Blended Supplementary Cementitious Materials, 5 pp. ASTM, USA (2021)
23. ASTM C1778: Standard Guide for Reducing the Risk of Deleterious Alkali-Aggregate Reaction in Concrete, 13 pp. ASTM, USA (2022)
24. ASTM C1797: Standard Specification for Ground Calcium Carbonate and Aggregate Mineral Fillers for Use in Hydraulic Cement Concrete, 6 pp. ASTM, USA (2018)
25. ASTM C1866/C1866M: Standard Specification for Ground-Glass Pozzolan for Use in Concrete, 5 pp. ASTM, USA (2020)
26. Bérubé, M.A., Fournier, B.: Alkalis releasable by aggregates in concrete: significance and test methods. In: Proceedings of the 12th International Congress on Alkali-Aggregate Reactions in Concrete, Beijing, China, pp. 17–30 (2004)
27. Bérubé, M.A., Duchesne, J., Dorion, J.F., Rivest, M.: Labouratory assessment of alkali contribution by aggregates to concrete and application to concrete structures affected by alkali-silica reactivity. Cem. Concr. Res.. Concr. Res. **32**(8), 1215–1227 (2002)
28. BRE Digest 330: Alkali-Silica Reaction in Concrete. Background to the Guidance Notes, 8 pp. BRE, UK (2004)
29. BS 812-123: Testing Aggregates - Part 123: Method for Determination of Alkali-Silica Reactivity - Concrete Prism Method, 19 pp. BSI, UK (1999)
30. BS 1881-124: Testing Concrete. Methods for Analysis of Hardened Concrete, 46 pp. BSI, UK (2015)
31. BS 3892-1: Pulverized-Fuel Ash. Specification for Pulverized-Fuel Ash for Use as a Cementitious Component in Structural Concrete, 22 pp. BSI, UK (1982)
32. BS 6068-2.42: Water Quality. Physical, Chemical and Biochemical Methods Determination of Sodium and Potassium: Determination of Sodium by Atomic Absorption Spectrometry, 10 pp. BSI, UK (1993)
33. BS 6068-2.43: Water Quality. Physical, Chemical and Biochemical Methods Determination of Sodium and Potassium: Determination of Potassium by Atomic Absorption Spectrometry, 10 pp. BSI, UK (1993)
34. BS 6068-2.44: Water Quality. Physical, Chemical and Biochemical Methods Determination of Sodium and Potassium: Determination of Sodium and Potassium by Flame Emission Spectrometry, 10 pp. BSI, UK (1993)
35. BS 8500-1:2015+A2:2019: Concrete. Complementary British Standard to BS EN 206. Method of Specifying and Guidance for the Specifier, 72 pp. BSI, UK (2020)
36. BS 8500-2:2015+A2:2019: Concrete. Complementary British Standard to BS EN 206. Specification for constituent materials and concrete, 52 pp. BSI, UK (2019)
37. BS EN 196-2: Method of Testing Cement Chemical Analysis of Cement, 78 pp. BSI, UK (2013)
38. CEN CR 1901: Regional Specifications and Recommendations for the Avoidance of Alkali-Silica Reactions in Concrete, 63 pp. CEN (2000)
39. Constantiner, D., Diamond, S.: Pore solution analysis: are there pressure effects? In: Mechanisms of Chemical Degradation of Cement-Based Systems, London. ISBN: 0419215700 (1997)
40. CSA A23.2-27A: Standard Practice to Identify Degree of Alkali-Aggregate Reactivity of Aggregates and to Identify Measures to Avoid Deleterious Expansion in Concrete, pp. 260–269. Canadian Standards Association, Canada (2019)
41. CSA A864-00: Guide to the Evaluation and Management of Concrete Structures Affected by Alkali-Aggregate Reaction, 121 pp. Canadian Standards Association, Canada (2005)

42. CSA A3001: Cementitious Materials for Use in Concrete, 253 pp. Canadian Standards Association, Canada (2018)
43. EN 196-2: Method of Testing Cement. Chemical Analysis of Cement, 73 pp. CEN, Brussels (2013)
44. EN 197-1: Composition, Specifications and Conformity Criteria for Common Cements, 56 pp. CEN, Brussels (2011)
45. EN 206: Concrete. Specification, Performance, Production and Conformity, 93 pp. CEN, Brussels (2013)
46. EN 450-1: Fly Ash for Concrete - Part 1: Definition, Specifications and Conformity Criteria, 30 pp. CEN, Brussels (2013)
47. EN 480-12: Admixtures for Concrete, Mortar and Grout. Test Methods. Determination of the Alkali Content of Admixtures, 9 pp. CEN, Brussels (2005)
48. EN 934-1: Admixtures for Concrete, Mortar and Grout. Common Requirements, 10 pp. CEN, Brussels (2008)
49. EN 934-2:2009+A1:2012: Admixtures for Concrete, Mortar and Grout—Part 2: Concrete Admixtures - Definitions, Requirements, Conformity, Marking and Labelling, 24 pp. CEN, Brussels (2012)
50. EN 1008: Mixing Water for Concrete. Specification for Sampling, Testing and Assessing the Suitability of Water, Including Water Recovered from Processes in the Concrete Industry, as Mixing Water for Concrete, 18 pp. CEN, Brussels (2003)
51. EN 1744-1:2009+A1:2012: Tests for Chemical Properties of Aggregates - Part 1: Chemical Analysis, 62 pp. CEN, Brussels (2012)
52. EN 12620:2002+A1:2008: Aggregates for Concrete, 56 pp. CEN, Brussels (2008)
53. EN 12878: Pigments for the Colouring of Building Materials Based on Cement and/or Lime. Specifications and Methods of Test, 30 pp. CEN, Brussels (2014)
54. EN 13055-1:2002: Lightweight Aggregates - Part 1: Lightweight Aggregates for Concrete, Mortar and Grout, 36 pp. CEN, Brussels (2002)
55. EN 13263-1:2005+A1:2009: Silica Fume for Concrete - Part 1: Definitions, Requirements and Conformity Criteria, 23 pp. CEN, Brussels (2009)
56. EN 14216: Cement. Composition, Specifications and Conformity Criteria for Very Low Heat Special Cements, 24 pp. CEN, Brussels (2015)
57. EN 15167-1: Ground Granulated Blast Furnace Slag for Use in Concrete, Mortar and Grout - Part 1: Definitions, Specifications and Conformity Criteria, 20 pp. CEN, Brussels (2006)
58. EN 15743:2010+A1:2015: Supersulphated Cement - Composition, Specifications and Conformity Criteria, 27 pp. CEN, Brussels (2015)
59. FD P18-464: Concrete - Provisions for the Prevention of Alkali Silica Reactions, 25 pp. AFNOR, France (2021)
60. FD P18-541: Aggregates - Guide for the Drafting of the Quarry Specification for the Purposes of the Prevention of Deleterious Effects of Alkali-Aggregate Reactions, 14 pp. AFNOR, France (2015)
61. FD P18-542: Agregates - Criteria for Qualification Natural Agregates for Hydraulic Concrete with Respect to the Alkali-Reaction, 18 pp. AFNOR, France (2017)
62. Fernández-Jiménez, A., Puertas, F.: The alkali-silica reaction in alkali-activated granulated slag mortars with reactive aggregates. Cem. Concr. Res.. Concr. Res. **32**(7), 1019–1024 (2002)
63. Gifford, P.M., Gillott, J.E.: Alkali-silica reaction (ASR) and alkali-carbonate reaction (ACR) in activated blast furnace slag cement (ABFSC) concrete. Cem. Concr. Res.. Concr. Res. **26**(1), 21–26 (1996)
64. Grandubé: Grandeurs Associeés À La Durabilité Des Bétons, Arliguie and Hornain Cordinators, 438 pp. Presses de l'école nationale des Ponts et Chaussées, France (2007)
65. ISO 22965-1: Concrete - Part 1: Methods of Specifying and Guidance for the Specifier, 12 pp. ISO, Switzerland (2007)
66. ISO 22965-2: Concrete - Part 2: Specification of Constituent Materials, Production of Concrete and Compliance of Concrete, 41 pp. ISO, Switzerland (2007)

67. ISO 29581-1: Cement. Test Methods. Analysis by Wet Chemistry, 49 pp. ISO, Switzerland (2009)
68. ISO 29581-2: Cement. Test Methods. Chemical Analysis by X-ray Fluorescence, 30 pp. ISO, Switzerland (2010)
69. IStructE: Structural Effects of Alkali-Silica Reaction, 48 pp. IstrucE Ltd. (1992)
70. JIS A 1145: Method of Test for Alkali-Silica Reactivity of Aggregates by Chemical Method, 16 pp. Japanese Standards Association (2017)
71. JIS A 1146: Method of Test for Alkali-Silica Reactivity of Aggregates by Mortar-Bar Method, 13 pp. Japanese Standards Association (2021)
72. JIS A 5308: Ready-Mixed Concrete, 83 pp. Japanese Standards Association (2019)
73. JIS A 6201: Fly Ash for Use in Concrete, 28 pp. Japanese Standards Association (2019)
74. JIS A 6204: Chemical Admixtures for Concrete, 32 pp. Japanese Standards Association (2021)
75. JIS A 6206: Ground Granulated Blast-Furnace Slag for Concrete, 16 pp. Japanese Standards Association (2013)
76. JIS R 5202: Methods for Chemical Analysis of Cements, 109 pp. Japanese Standards Association (2019)
77. JIS R 5204: Chemical Analysis Method of Cement by X-ray Fluorescence, 54 pp. Japanese Standards Association (2019)
78. JIS R 5210: Portland Cement, 20 pp. Japanese Standards Association (2013)
79. JIS R 5211: Portland Blast-Furnace Slag Cement, 14 pp. Japanese Standards Association (2014)
80. JIS R 5213: Portland Fly Ash Cement, 12 pp. Japanese Standards Association (2019)
81. Lane, D.S., Ozyildirum, H.C.: Evaluation of effect of Portland cement alkali content, fly ash, ground slag, and silica fume on alkali-silica reactivity. Cem. Concr. Aggreg. **21**(2), 126–140 (1999)
82. LNEC E 461: Concrete. Methodologies to Prevent Expansive Chemical Reactions of Internal Origin, 12 pp. LNEC (in Portuguese) (2021)
83. LNEC MI 115: Concrete. Determination of the Sulphate, Chloride and Alkali Content. LNEC. (in Portuguese) (2007)
84. Lu, D., Fournier, B., Grattan-Bellew, P.E.: Evaluation of accelerated test methods for determining alkali-silica reactivity of concrete aggregates. Cem. Concr. Compos. **28**, 546–554 (2006)
85. Massazza, F.: Pozzolana and Pozzolanic Cements. In: Hewlett, P.C. (ed.) Lea's Chemistry of Cement and Concrete, 4th ed., pp. 471–631. Elsevier Science & Technology Books (2004)
86. Menéndez, E., Silva, A.S., Duchesne, J.: Recommendation of RILEM TC 258-AAA: RILEM AAR-8: determination of potential releasable alkalis by aggregates in concrete. Mater. Struct. **54**, 205 (2021). https://doi.org/10.1617/s11527-021-01682-1
87. NF EN 196-2: Methods of Testing Cement—Part 2: Chemical Analysis of Cement, 80 pp. AFNOR, France (2013)
88. NF EN 480-12: Admixtures for Concrete, Mortar and Grout - Test Methods - Part 12: Determination of the Alkali Content of Admixtures, 10 pp. AFNOR, France (2006)
89. NF P18-454: Concrete - Reactivity of a Concrete Formula with Regard to the Alkali-Aggregate Reaction—Performance Test, 27 pp. AFNOR, France (2021)
90. NF P 18-545: Aggregates - Defining Elements, Conformity and Coding, 74 pp. AFNOR, France (2011)
91. NP 1382: Aggregates for Mortars and Concretes. Determination of Soluble Alkali Content. Flame Spectrophotometry Process. IPQ, Caparica (1976)
92. PD 6682-1: Aggregates. Aggregates for Concrete. Guidance on the Use of BS EN 12620, 22 pp. CEN, Brussels (2003)
93. PD CEN/TR 16349: Framework for a Specification on the Avoidance of a Damaging Alkali-Silica Reaction (ASR) in concrete, 14 pp. CEN, Brussels (2012)
94. Pontes, J., Santos Silva, A., Faria, P.: Evaluation of pozzolanic reactivity of artificial pozzolans. Mater. Sci. Forum **730–732**, 433–438 (2013)

95. Quarcioni, V.A., Chotoli, F.F., Coelho, A.C.V., Cincotto, M.A.: Indirect and direct Chapelle's methods for the determination of lime consumption in pozzolanic materials. Revista IBRACON de Estruturas e Materiais **8**, 1–7 (2015)
96. Ramlochan, T., Thomas, M., Gruber, K.A.: The effect of metakaolin on alkali–silica reaction in concrete. Cem. Concr. Res.. Concr. Res. **30**, 339–344 (2000)
97. RILEM AAR-7.1: International specification to minimise damage from alkali reactions in concrete - Part 1: alkali-silica reaction. In: Nixon, P., Sims, I. (eds.) RILEM Recommendations for the Prevention of Damage by Alkali-Aggregate Reactions in New Concrete Structures. RILEM State-of-the-Art Reports, vol. 17. Springer, Dordrecht (2016). https://doi.org/10.1007/978-94-017-7252-5_8
98. RILEM AAR-7.3: Preliminary International Specification to Minimise Damage from Alkali-Aggregate Reactions in Concrete - Part 3: Concrete Dams and Other Hydro Structures, In: Nixon, P., Sims, I. (eds.) RILEM Recommendations for the Prevention of Damage by Alkali-Aggregate Reactions in New Concrete Structures. RILEM State-of-the-Art Reports, vol. 17. Springer, Dordrecht (2016). https://doi.org/10.1007/978-94-017-7252-5_10
99. Sabir, B.B., Wild, S., Bai, J.: Metakaolin and calcined clays as pozzolans for concrete: a review. Cement Concr. Compos.Concr. Compos. **23**(6), 441–454 (2001)
100. Shafaatian, S.M.H., Akhavan, A., Maraghechi, H., Rajabipour, F.: How does fly ash mitigate alkali–silica reaction (ASR) in accelerated mortar bar test (ASTM C1567)? Cement Concr. Compos.Concr. Compos. **37**, 143–153 (2013)
101. Shehata, M.H., Thomas, M.D.A.: Use of ternary blends containing silica fume and fly ash to suppress expansion due to alkali-silica reaction in concrete. Cem. Concr. Res.. Concr. Res. **32**(3), 341–349 (2002)
102. Soares, D., Santos Silva, A., Mirão, J., Ramos, V., Fernandes, I, Menéndez, E.: Assessment of alkalis released by aggregates. Contribution to the alkalinity increase and AAR development in concrete. In: Proceedings of the second International Dam World Conference, Lisbon, 10 pp. LNEC (2015)
103. Soares, D., Santos Silva, A., Mirão, J., Fernandes, I., Menéndez, E.: Study on the factors affecting alkalis release from aggregates into ASR. In: Proceedings of the 15th International Conference on Alkali-Aggregate Reaction in Concrete, São Paulo, Brazil, 10 pp. (2016)
104. Touzé, P., Godart, B.: Évaluation de la teneur en alcalins équivalents actifs dans les ciments: Méthode d'essai LPC n° 48, 4 pp. IFSTTAR, France (1997)
105. TR 30: Alkali-Silica Reaction. Minimizing the Risk of Damage to Concrete, 72 pp. Concrete Society, UK (1999)
106. Waller, V., Naproux, P., Larrard, F.: Contribution des fumées de silice et des cendres volantes silico-alumineuses à la résistance en compression du béton - Quantification. Bulletin des Labouratoires des Ponts et Chaussées **208**, 53-65 (1997)
107. Wild, S., Khatib, J.M.: Portlandite consumption in metakaolin cement pastes and mortars. Cem. Concr. Res.. Concr. Res. **27**(1), 137–146 (1997)
108. XP P 18-544: Aggregates - Determination of Active Lime Water-Soluble Alkalis, 13 pp. AFNOR, France (2015)

Annex: Selected Examples of Concrete Structures in Which the Development of ASR Was Affected by Alkali Distribution

Colin Giebson, Leandro Sanchez, and Esperanza Menéndez Méndez

Abstract The examples in this annex highlight the importance of considering external sources (A.2: Pavements) and internal long-term sources (A.3: Dams) of alkalis. ASR in concrete pavements (A.2) was a concern in Germany for many years before it was shown that externally applied NaCl as de-icing agent during winter maintenance action triggered ASR in many cases. At that time, this impact was not considered at all by the standard test methods as the German fog chamber test. A novel test method, the climate simulation concrete prism test, clearly showed that some aggregates quickly respond to NaCl, even in combination with low-alkali cements. New test protocols and regulations adopted gradually since 2005 in Germany reduced the ASR damage to concrete pavements greatly. For dams (A.3), on the other hand, alkalis from certain aggregate constituents as feldspars or SCMs are slowly released and may contribute to ASR in the long-term. Often, the serviceability is affected considerably as the loss of spillway gate clearances. Since the replacement of ASR-affected dams is usually not an option, sophisticated, time-consuming and expensive measures as slot cutting or membrane sealings are necessary to maintain their serviceability. Hence, the RILEM AAR-7.3 document provides extended guidance in how to avoid ASR in hydraulic structures as the use of aggregates with sufficiently low amounts of releasable alkalis, the use of SCMs but also special design and construction measures.

C. Giebson (✉)
Bauhaus-Universität Weimar, Weimar, Germany
e-mail: colin.giebson@uni-weimar.de

L. Sanchez
University of Ottawa, Ottawa, Canada
e-mail: leandro.sanchez@uottawa.ca

E. Menéndez Méndez (Deceased)
Institute Eduardo Torroja of Construction Science (IETcc-CSIC), C/Serrano Galvache, 28033 Madrid, Spain
e-mail: emm@ietcc.csic.es

E. Menéndez Méndez and A. Santos Silva (eds.), *Total Alkalis in Concrete—Contribution to Alkali-silica Reaction*, RILEM State-of-the-Art Reports 40, https://doi.org/10.1007/978-3-031-56164-1

Keywords External alkalis · Releasable alkalis · Pavements · Dams · Climate simulation concrete prism test · RILEM AAR-7.3

A.1 Introduction

Alkalis in Portland cement (PC) are generally expressed as sodium oxide (NaO_2) and potassium oxide (K_2O). Initially, the alkalis are found within the anhydrous phases of PC, and then dissolve during the process of hydration into Na^+, K^+ and OH^- ions [5, 39]. These ions accumulate in the concrete pore solution under high CaO conditions, since they are not constituents of PC's hydration process [17].[1]

Alkali-silica reaction (ASR) is an acid–base reaction, where the "acid reactants" are silica-based phases from the aggregates used to fabricate concrete whereas the "base elements" are the sodium/potassium hydroxides present in the material's pore solution. This reaction generates a secondary product (i.e., calcium-sodium/potassium silicate hydrate) that swells upon moisture uptake, leading to expansion and damage of the affected material [16]. To be triggered and maintained, ASR needs a high pH environment [36], which is often generated due to the presence of the alkali hydroxides in the material's pore solution [20].

The alkalis of ASR-affected concrete are progressively consumed through the secondary products formation; however, when ASR-induced cracking (often generated within the aggregate particles) extends to the cement paste, an ion-exchange process takes place with Ca^+, which progressively recycles Na^+ and/or K^+ alkalis back to the pore solution [48].

Previous investigations as per [3, 29, 45] demonstrated that the alkali content of the concrete pore solution increases as a function of PC hydration, yet, it has also been found that the total alkali may also raise at later ages, likely due to the alkali's release of the aggregates. The latter suggests that aggregates bearing alkalis may increase the risk associated to ASR-initial and/or further induced development. Hence, determining the total and/or "available" alkali content in concrete (and in the aggregates) is essential to establish management protocols for appraising the potential of further induced expansion and damage and for selecting rehabilitation measures of ASR-affected structures [2, 4].

A wide number of researcher projects worldwide have been studying and measuring the alkali content of concrete in the laboratory [14]. Yet, the comparison amongst laboratory and field data along with the interpretation of laboratory test results for use on the decision making of ASR-affected infrastructure still requires further consideration [14, 15, 21, 28].

[1] This annex was written with the cooperation of Ricardo García-Roves.

Fig. A.1 Typical discolourations along the joints, especially at the cross points, of a concrete pavement indicating a beginning ASR

A.2 Pavements

Pavement concrete is one of the most stressed types of concrete. The placing must be done under given weather conditions, curing is delicate and in Central Europe, pavements are exposed to changing temperature and moisture conditions including freeze–thaw cycles and the application of de-icers. Moreover, the permanent traffic loads create pre-damages (micro cracks) that support the progress of deterioration processes [23]. In the late 1990s, an increasing number of ASR-damaged concrete highway pavements occurred in Germany although the existing guideline for preventing ASR in concrete, first released in 1974, was strictly followed [13]. Affected highway sections showed first signs of ASR after about 8–15 years in service, i.e., after half the intended service life of 30 years at the latest. The typical indication of a starting ASR in pavements are dark discolourations along the joints, especially at the cross points (Fig. A.1). Further on map cracking appears and the deterioration moves towards the centre of the slab (Fig. A.2). Between the appearance of the first signs (discolourations along the joints) and more clear indications (map cracking) several years may elapse. However, severe cracking and even spalling can occur subsequently within one more year only [6].

Fig. A.2 Map cracking at the cross points of a concrete pavement

Two damaged German highway pavement concretes, built in 1991/93, were examined by thin section analysis in 2006 and it was found that ASR was the cause for the distress. In concrete 1, the gravel and rhyolite aggregates were affected by ASR (Figs. A.3 and A.4) and the damage occurred after 12 years in service. In concrete 2, the granodiorite aggregates were affected by ASR (Figs. A.5 and A.6) and the damage occurred after 8 years only. Both concretes contained 360 kg/m^3 of Portland cement (CEM I, Na_2O_{eq} 0.90–0.95 wt%) and had a w/c of 0.40–0.45. The concretes were air-entrained with 4.5–5.0 vol% air. The objective of a subsequent study in [41, 42] was to compare the field experience for those two concretes with ASR performance testing by using the climate simulation concrete prism test (CS-CPT), explained in detail elsewhere [19, 43]. Concrete 1 was reproduced in the laboratory by using all the aggregates from the original deposits and a Portland cement (Na_2O_{eq} 0.90 wt%) from the original plant. For concrete 2, the granodiorite aggregates from the original quarry were used in a typical mixture for pavement concrete that meets the latest requirements at that time for highway pavements.

To evaluate if the geological situation in the deposits has changed compared to the situation 15 years ago, the coarse rhyolite aggregates from a core of concrete 1 were extracted for a mortar-bar test by means of shockwave crushing. The results showed that the rhyolite from the core is highly alkali reactive as well as the gravel

Fig. A.3 Highway pavement (concrete 1) after about 14 years in service

and the rhyolite from the new batches (Fig. A.7). The German fog chamber test, however, identified the rhyolite as non-reactive (Fig. A.8). Contrary, the CS-CPT showed deleterious expansion after exposing the prisms for 7 cycles (21 weeks) to NaCl solution (Fig. A.9) what finally corresponds to the field experience. After some more years and repairs, the deterioration for concrete 1 in the field had further increased, resulting in its replacement in 2011, i.e., after about 20 years in service only.

For concrete 2, the granodiorite was assumed to be non-reactive according to the mortar-bar test and the German fog chamber test as well (Figs. A.10 and A.11) but deleterious expansion occurred in the CS-CPT after exposing the prisms for 8 cycles (24 weeks) to NaCl solution (Fig. A.12). A subsequent thin section analysis provided clear evidence for an ASR (cracks, ASR-gel), triggered by the granodiorite (Fig. A.13), what clearly corresponds to the field experience.

In 2009, an evaluation revealed that about 10% of the nearly 4000 km concrete highway pavements in Germany are either proved or suspected to suffer from ASR [12]. The high amount of highway pavement concrete (nearly 50%) on the total number of ASR-damaged structures led to the suspicion that special mechanisms promote the ASR in pavements. Studies (e.g., [22, 40, 46, 47]) revealed that an application of NaCl de-icer solution on concrete prisms with alkali-reactive aggregates highly accelerate the ASR (Sect. 2.2.5). Since 2004, several hundred concrete mixtures comprised mainly of field representative concrete with different types of cement and mostly slow/late type alkali-reactive aggregates have been tested with the CS-CPT. The lessons learned are that ASR can be clearly triggered and accelerated considerably in concrete with alkali-reactive aggregates and exposed to alkali-containing de-icers. The alkali-reactivity of some aggregates is assessed incorrectly by mortar-bar tests and the German fog chamber test as well. Petrography provides powerful supplementary information about the aggregates but is often not enough for a conclusive assessment. So far, the CS-CPT provides results that are in good agreement with the field experience for pavement concrete and is part of the German guideline for building highway pavements since 2018 [19].

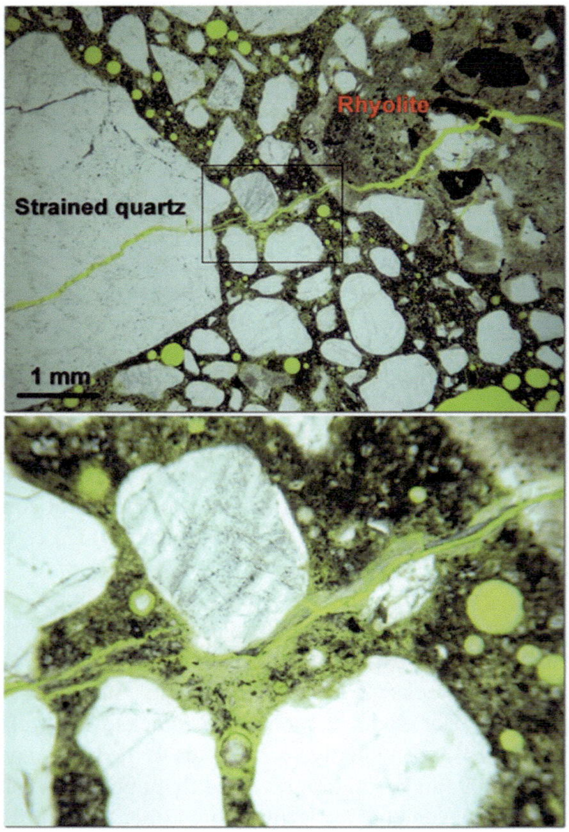

Fig. A.4 Thin section image of concrete 1 (14 years old) showing cracks in aggregate grains, partially filled with ASR gel

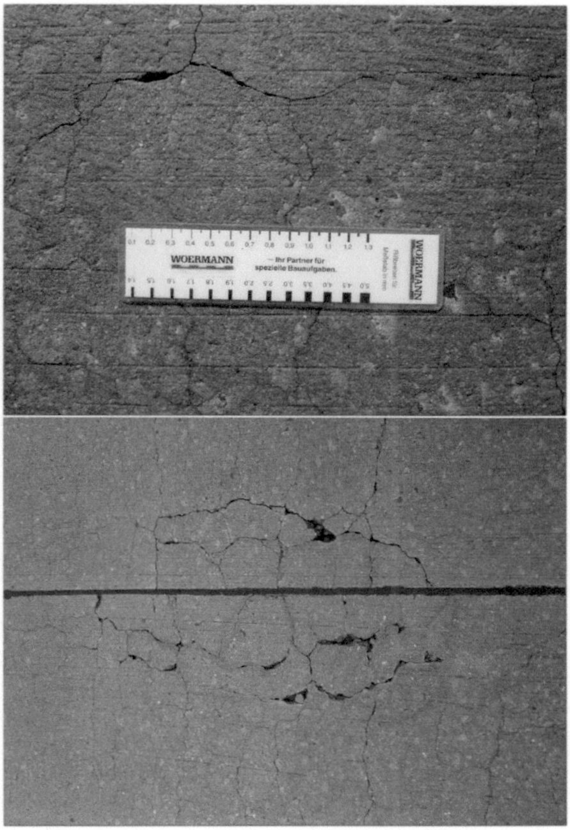

Fig. A.5 Highway pavement (concrete 2) after about 9 years in service

A.3 Dams

Dams are unique structures in relation to expansive reactions and especially in relation to the ASR reaction, this is due to various factors, such as:

- Large volumes of concrete;
- The aggregates that necessarily have to be local and can be heterogeneous;
- The different stress states to which various areas of the dam are subjected;
- The morphology of the dam (arch, gravity, buttress, arch-gravity, …);
- The different environmental conditions to which different areas are subjected (insolation, wind, temperature, etc.);
- The different conditions in which water and humidity are found in different areas of the concrete: filtering water, tide run, saturated concrete, areas of surface evaporation, etc.

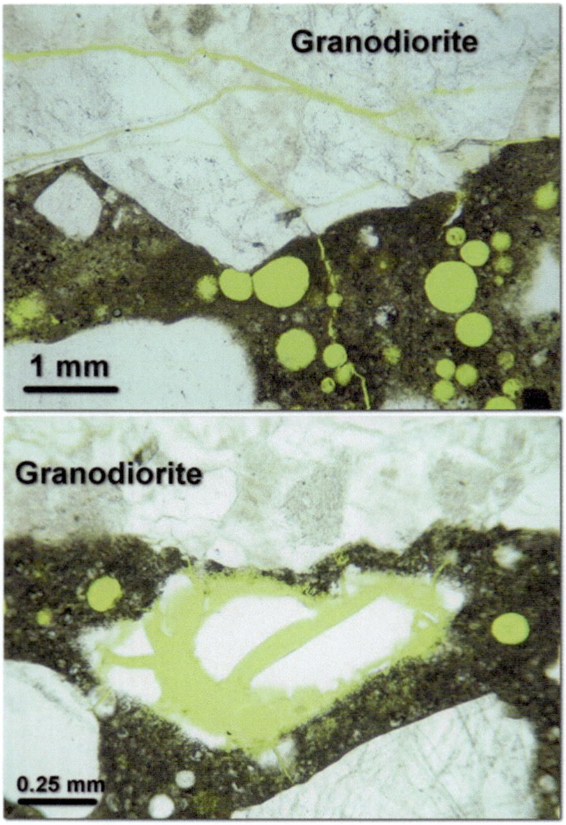

Fig. A.6 Thin section image of concrete 2 (9 years old) showing pores with ASR gel next to a granodiorite grain

A.3.1 Swelling in Dams

With respect to the internal swelling reactions most attention has been paid to alkali-aggregate reactions as a cause of swelling of concrete in dams.

But there are two groups of underlying mechanisms causing internal swelling in dams:

- *Alkali-Aggregate Reactions (AAR)*: including alkali-silica reaction (ASR) and alkali-carbonate reaction (ACR); and
- *Sulfate related deteriorations*: including Internal Sulfate Attack (ISA) including pyrite oxidation, External Sulfate Attack (ESA), thaumasite formation and Delayed Ettringite Formation (DEF).

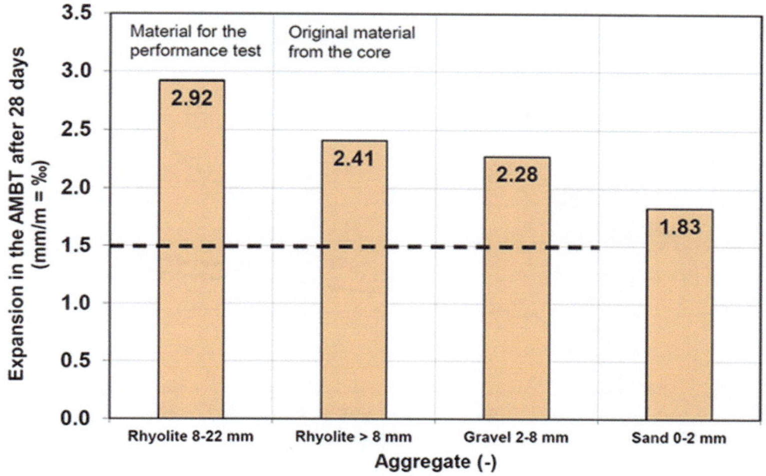

Fig. A.7 Mortar-bar test results for the aggregates used in concrete 1

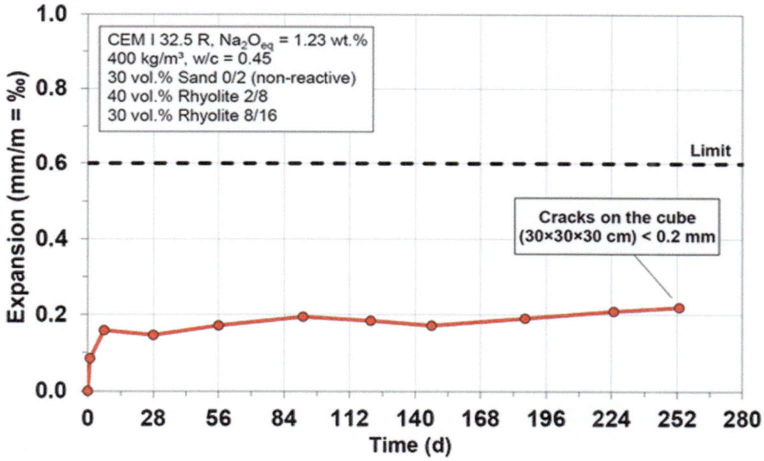

Fig. A.8 German fog chamber test result for the rhyolite used in concrete 1

It is also now known that the phenomenon may continue for longer periods of time than initially thought. Given the pervasiveness of the expansion problem and the realization that most dams will need to continue to operate, well beyond the 30–50 year service life often initially considered, long-term performance issues are becoming more and more important.

It is important to note that several of the large dams exhibiting progressive expansion reactions, most commonly ASR varieties, met testing criteria at the time of their construction that are generally expected to prevent such behaviour.

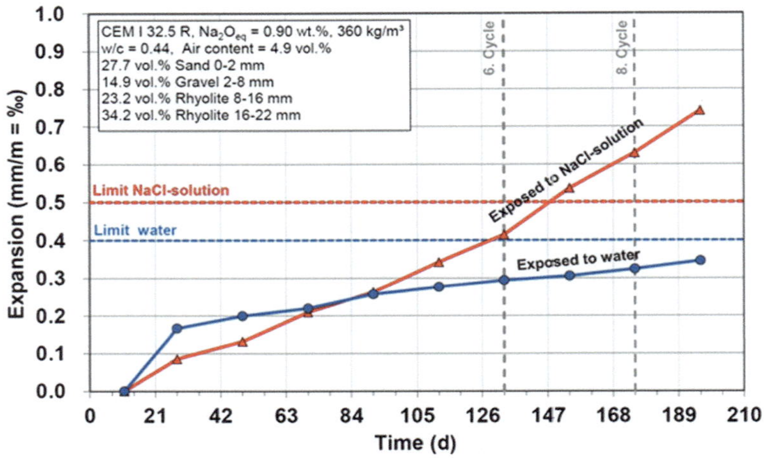

Fig. A.9 Expansions during the CS-CPT for concrete 1, exposed to NaCl solution and water (control)

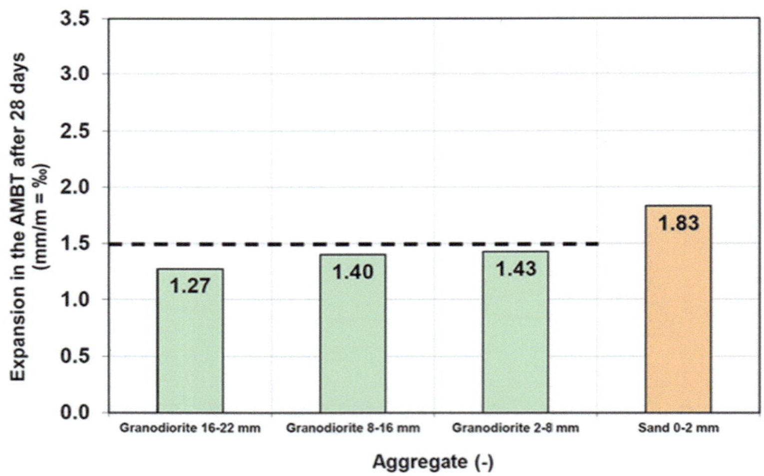

Fig. A.10 Mortar Mortar-bar test results for the aggregate used in concrete 2

A.3.2 Long Term Challenge

With the time have been detected other long-term sources of "unscheduled" alkalis, including:

- releasable alkalis from aggregates;
- releasable alkalis from additions/supplementary cementitious materials;

- recycling of alkalis from matured gel.

If present, these have the potential to cause expansion and deterioration to continue for a very long time.

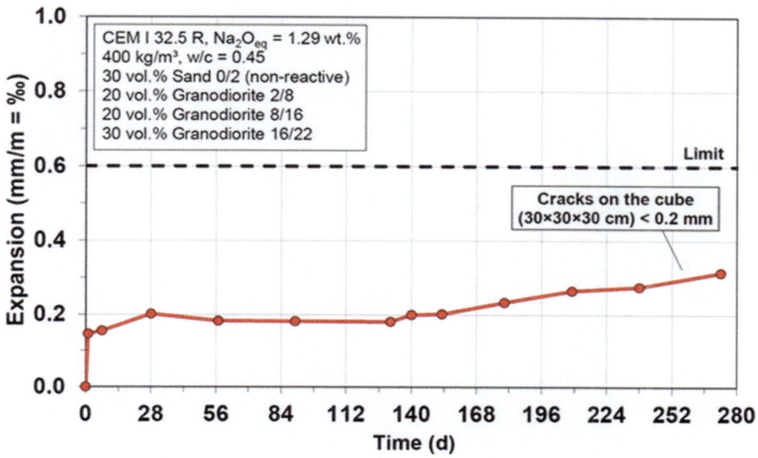

Fig. A.11 German fog chamber test result for the granodiorite used in concrete 2

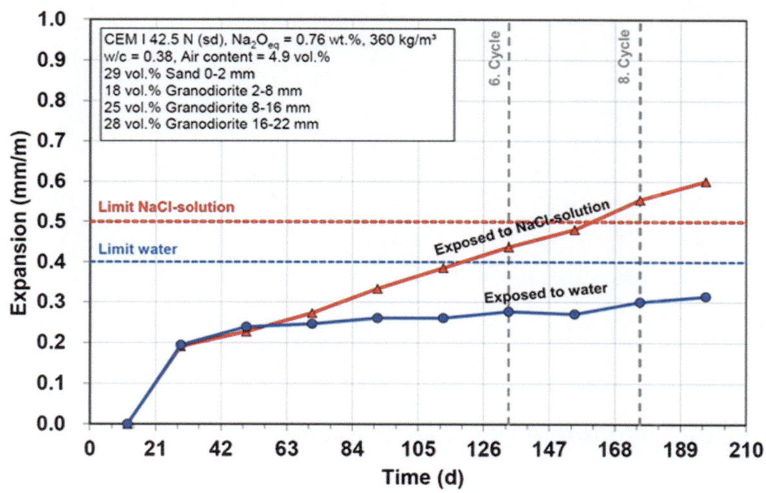

Fig. A.12 Expansions during the CS-CPT for concrete 2, exposed to NaCl solution and water (control)

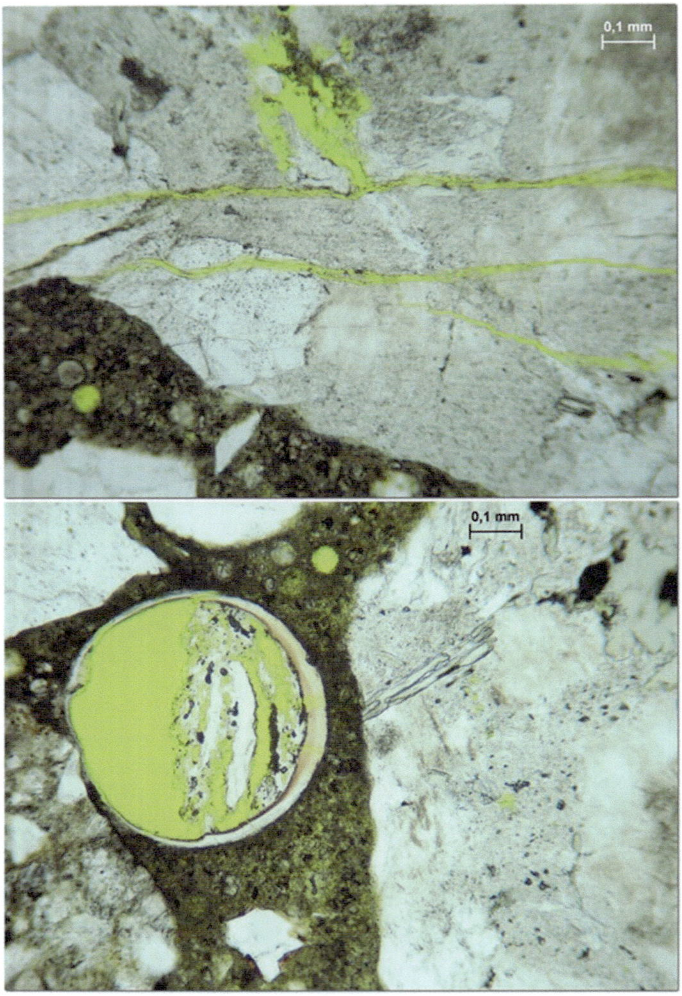

Fig. A.13 Thin section image of concrete 2 after the CS-CPT (NaCl) shows a granodiorite grain (G) with micro cracks

A.3.3 Confirmation of the Alkalis Release

For many years, the possibility that alkalis may be released from feldspars and other alkali-bearing aggregate components directly into concrete pore solutions has been explored by various mostly indirect methods.

Constantiner and Diamond in 2003 [11] demonstrated the effect of the alkali release by direct measurements of the concentrations of alkalis in pore solutions expressed from mortars by high pressure with and without such aggregates. These findings lend further emphasis to the importance of alkali release in many practical ASR problems. In addition, Menéndez et al. [30] in the ICAAR 2012 shows a direct relationship between the presence of Na-feldspars and K-feldspars with the formation of ASR rich in sodium or potassium, respectively, in a Spanish dam with granitic aggregates. This was done using microstructural analysis. Also, Soares et al. [44] mention the contribution of alkali leaching from aggregates to the increase of ASR in concrete.

A.3.4 Cases of Alkali-Silica Reaction in Dams—ICOLD

The first case was detected in Parker Dam in Arizona was found to have ASR in 1940. In 1991 ICOLD Bulletin 79, Alkali-Aggregate Reaction in Dams discussed the problem of alkali-aggregate reactions (AAR) in dams and referred to a number of cases worldwide [25]. In 1992 over 100 large dams and hydroelectric projects were identified to be seriously affected by ASR impacting dam safety and operations at USCOLD in 1995 [52]. In 2007 experience was reviewed at the 2007 ICOLD/ SPANCOLD Workshop [7] on Expanding Concrete in Dams held in Granada, Spain. And, in other papers in ICOLD Symposium [8], Hydropower and Dams [8], Symposium A, Long-term behaviour of dams [26]; International Workshop on Managing AAR in Dams [10], Spanish experience with AAR & ISA [31], ICOLD Bulletin 184: Managing Expansive Chemical Reactions in Concrete Dams [27].

In the International Workshop on Managing AAR in Dams, ICOLD 2013 Post Tour & Workshop B4–South-eastern US Dams with Alkali-Aggregate Reactions, August 20, 2013 ICOLD-Annual Meeting 2013 Seattle, WA, USA, Robin Charlwood and others presented different experiences about alkali-silica reactions [10].

A.3.4.1 Cases by Typology of Dams

Gravity Dams:

Expansion effects have caused concern particularly at changes of geometry; for instance, at the change from a straight section to a smaller curved abutment section and also around galleries creating cracks that may adversely affect sliding stability [10].

An example of this type of dams is Mactaquac dam, in Canada. Their characteristics are presented in Table A.1, and some images of the structure and measured expansion are shown in Figs. A.14, A.15 and A.16 (Fletcher and Curtis in Charlwood [10]).

In Fig. A.14a, b are shown the aspect of the Mactaquac dam, its symtoms of damage and the distributions of new project with the commonly (a) and abbreviations used for various structures and slot locations (b). In Fig. A.15 is shown the AAR expansion at Mactaquac GS (John Fletcher, NB Power and Dam Curtis, Hatch;

Table A.1 Characteristics of Mactaquac gravity dam

Official name	Mactaquac Generating Station
Country	**Canada**
Location	**Bright Parish/Kingsclear Parish, York County, New Brunswick**
Coordinates	45°57′15.2″N 66°52′5.8″W
Status	Operational
Construction began	1965
Opening date	1968
Owner(s)	NB Power
Type of dam	**Embankment dam**
Impounds	**Saint John River**
Height	55 m (180 ft)
Length	518 m (1699 ft)
Spillways	2
Spillway type	Concrete
Spillway capacity	16282.19 m^3 (575000 cu ft) per second
Reservoir	
Creates	Mactaquac Headpond
Total capacity	1307491000 m3 (4.61736 × 1010 cu ft)
Surface area	83.85 km^2 (20720 acres)
Maximum water depth	39.624 m (130.00 ft)
Power Station	
Type	**Run-of-the-river**
Hydraulic head	31.7 m (104 ft) - 35.4 m (116 ft)
Turbines	6 × Kaplan
Installed capacity	653 MW
Annual generation	1600 GWh (5800 TJ)

August 20, 2013). And, in Fig. A.16 cutting with diamond wire sawn in Mactaquac in 1988. In 2013 the prediction is done a recutting to increase the service life of the structure.

Arch Dams:

There are several cases of substantial upstream and vertical deformations, some reaching about 15 cm, sometimes resulting in diagonal cracks near the abutments and damage to horizontal lift joints which may affect stability.

An example of this type of dams is Stewart Mountain Arch Dam, some characteristics are presented in Table A.2 and some images of the structure are shown in Fig. A.17 (Fletcher and Curtis in Charlwood [10]).

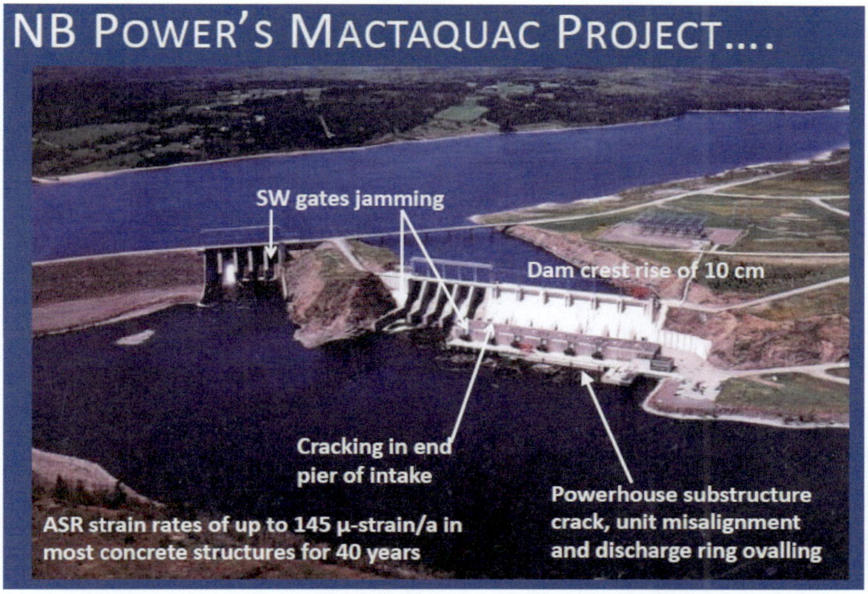

(a) NB Power´s Mactaquac project with objective of increasing the durability of the dam.

Commonly Used Abbreviations for Various Structures and Slot Locations

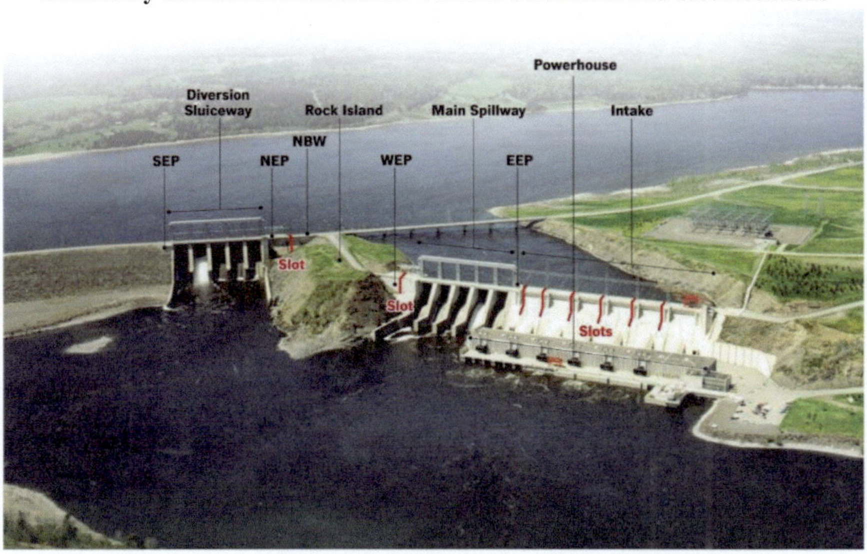

(b)

Fig. A.14 Dam of Mactaquac, in Canada

John Fletcher, NB Power and Dam Curtis, Hatch; August 20, 2013; Mananing AAR at Mactaquac GS

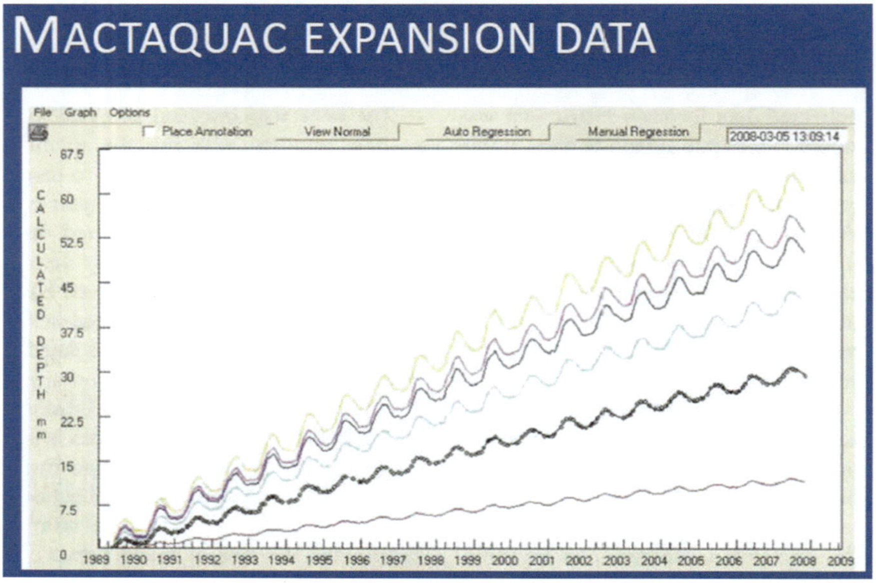

Fig. A.15 Expansion data from 1989 to 2009. Some blocks have much more expansion than ones

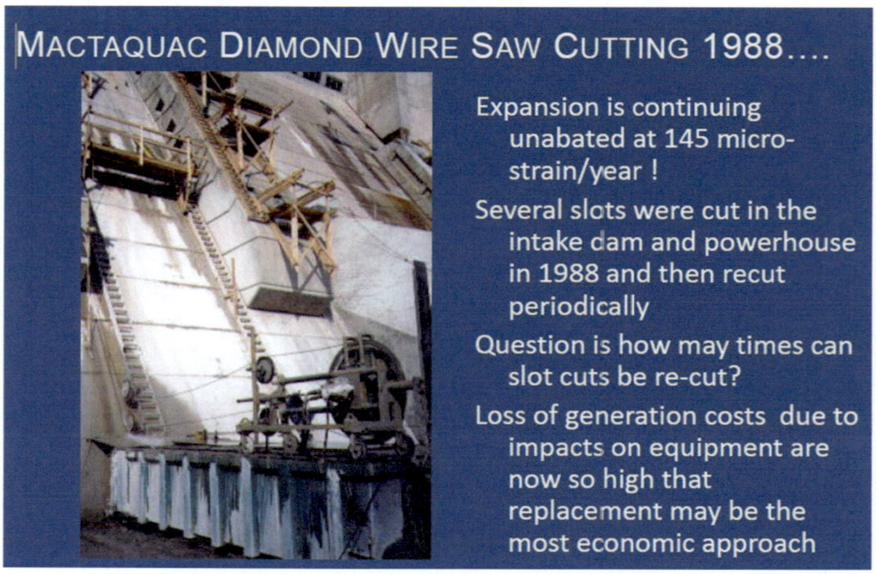

Fig. A.16 Characteristics of the expansion and repair by slot cuts and be re-cut

Fig. A.17 Stewart Mountain arch dam, in USA

Table A.2 Characteristics of Stewart Mountain Arch dam

Country	**United States of America**
Location	**Maricopa County, Arizona**
Coordinates	33°34′00″N 111°32′08″W
Status	Operational
Construction began	1928
Opening date	1930
Owner(s)	**United States Bureau of Reclamation**
Type of dam	**Arch dam**
Impounds	**Salt River**
Height	207 feet (63 m)
Length	1260 feet (384 m)
Elevation at crest	1533 feet (467 m)
Width (crest)	8 feet (2.4 m)
Width (base)	33 feet (10 m)
Dam volume	120000 cubic yards (92000 m^3)
Reservoir	
Total capacity	73613 acre-feet (90800309 m^3)
Catchment area	6330 square miles (16400 km^2)
Surface area	1260 acres (510 ha)

(continued)

Table A.2 (continued)

Country	**United States of America**
Power Station	
Operator(s)	**Salt River Project**
Type	Conventional
Hydraulic head	116 feet (35 m)
Installed capacity	13 MW

Another example is Cahora Bassa dam (Mozambique). Some information about the dam and the studies developed are summarized below:

- ASR identified at Cahora Bassa in LNEC report in May 1995;
- In 2001 LNEC carried out an assessment of the concrete and make a finite-element model (FEM) structural analysis of the dam body;
- Most of this vertical expansion occurred after 1983. The average vertical expansion rate from 1983 to 2001 was estimated at 1 mm/year, maybe 15 micro-strain/year;
- The distribution of the mix design and alkalis is important in this case;
- An analysis to seek to assess the potential for future expansion was carried out by LNEC in 2003 which showed potential for continuation;
- The earlier levelling deformation data are unclear regarding signs of a slow-down in the rate of expansion;
- New high accuracy vertical extensometers have been installed to check current rates of expansion and check for any signs of a slow down in rate of ASR;
- No dam safety problems identified so far but there are issues related to spillway clearances;
- Now updating tests and FEM to try to better understand the future behaviour.

The main characteristics of Cahora Bassa dam are presented in Table A.3 and some images of the structure and the area of stresses are shown in Figs. A.18 and A.19 (Tembe, 2013, in Charlwood [10]).

Buttress:

Expansion effects in more complex geometries may be more severe.

An example of this type of dams is Pracana Dam (Portugal), the characteristics are presented in Table A.4 and some images of the structure are shown in Figs. A.20, A.21 and A.22 (Batista in Charlwood [10]).

Spillways and Gates:

Concrete movements have frequently caused problems with loss of spillway gate clearances and in some the gate guides are damaged thereby affecting the operability of spillway gates in high discharge conditions.

An example of spillway with damage by alkali-aggregate and repaired is the David D Terry Spillway of Hiwassee dam, their characteristics are summarized in Table A.5.

Table A.3 Main characteristics of Cahora Bassa dam, in Mozambique

Location	Mozambique
Coordinates	15°40′S 31°50′E
Lake type	**Reservoir**
Primary inflows	**Zambezi River**
Primary outflows	**Zambezi River**
Catchment area	56927 km^2 (21980 sq mi)
Basin countries	**Mozambique**
Max. length	292 km (181 mi)
Max. width	38 km (24 mi)
Surface area	2739 km^2 (1058 sq mi)
Average depth	20.9 m (69 ft)
Max. depth	157 m (515 ft)
Water volume	55.8 km^3 (45200000 acre·ft)
Surface elevation	314 m (1030 ft)

Fig. A.18 Aspect of the Cahora Bassa dam, in Mozambique

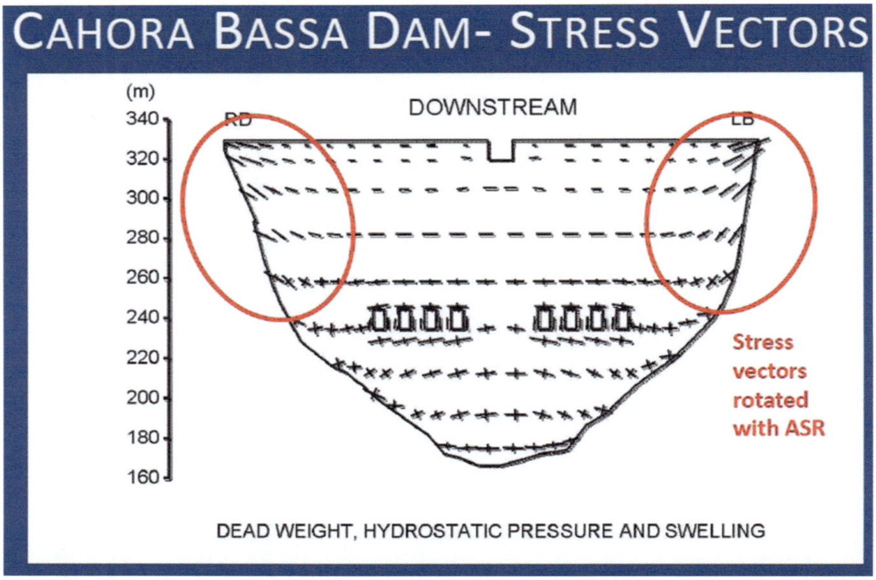

Fig. A.19 Area of stress in Cahora Bassa dam

Some images of the spillway are shown in Figs. A.23, A.24 and A.25 (Dodson in Charlwood [10]).

Powerhouses:

The most frequent effect of chemical expansion in powerhouses is "ovalling" of the concrete surrounding the units' thereby affecting runner and generator air gap clearances. In others misalignment of the units can occur.

A.3.5 General Aspects of the Evolution of the ASR – ICOLD

The evolution of the expansive reactions is uncertain, because of the complexity of the hydroelectric structures and the environmental aspects:

- Physical restraint is not a viable option due very large forces;
- Injection of lithium salts? Not viable to get penetration;
- Injection of CO_2, not viable to get penetration and unsafe. Environmental or synthetic CO_2 lowers the pH of the concrete due to its interaction with the portlandite and the alkalis of the porous phase. Although in the case of dams it is difficult to access the innermost parts of the structure due to the great thickness of the wall;

- Reduction of alkali content in pore fluid? Low diffusivity limits this to surfaces only;
- Drying to reduce RH (or degree of saturation) to <85%, sufficient moisture available internally after hydration;
- Draining large masses limited by low diffusivity;

Table A.4 Main characteristics of Pracana dam, in Portugal

Official name	Barragem de Pracana
Country	**Portugal**
Location	Municipality **Mação, Santarém District**
Coordinates	39°33′53.8″N 7°48′43.8″W
Purpose	Power
Status	Operational
Construction began	1947
Opening date	1950
Owner(s)	Companhia Portuguesa de Produção de Electricidade
Operator(s)	**Energias de Portugal**
Type of dam	**Concrete buttress dam**
Impounds	**Ocreza**
Height (foundation)	60 m (200 ft)
Length	245.5 m (805 ft)
Elevation at crest	115 m (377 ft)
Dam volume	144000 m^3 (5100000 cu ft)
Spillway type	Shaft and chute spillway
Spillway capacity	2560 m^3/s (90000 cu ft/s)
Reservoir	
Total capacity	111900000 m^3 (90700 acre·ft)
Active capacity	69000000 m^3 (56000 acre·ft)
Surface area	5.5 km^2 (2.1 sq mi)
Normal elevation	114 m (374 ft)
Operator(s)	**Energias de Portugal**
Commission date	1951
Hydraulic head	57 m (187 ft) (max)
Turbines	2 × 8 MW, 1 × 25 MW Francis-type
Installed capacity	41 MW
Annual generation	63.8 GWh

- Sealing using xylene sand/or membranes may allow drying over long periods but only for thin sections;
- Would need to limit water access from all directions;
- Sealing after drying during reservoir drawdown appears to have worked at Pracana Dam, but not normally possible.

Fig. A.20 Pracana dam **a** before the rehabilitation and **b** after the rehabilitation

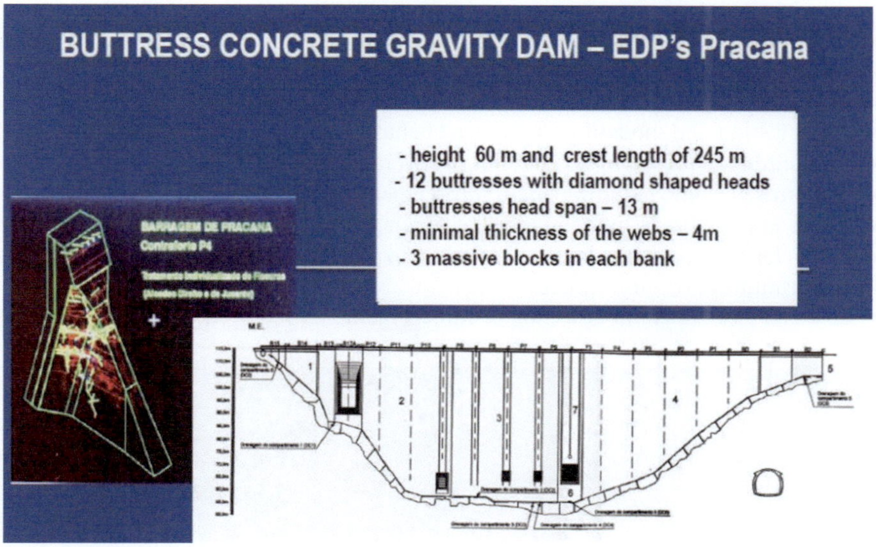

Fig. A.21 Pracana dam characteristics

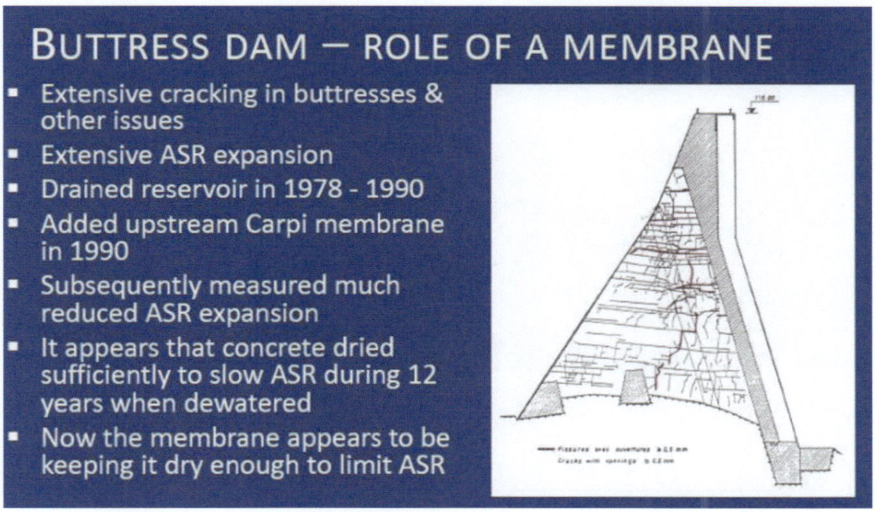

Fig. A.22 Pracana dam alteration of buttress

Some issues to managing on going alkali-silica reaction are the following:

- In some cases ASR will be exhausted after 30 years;
- In others ASR will continue indefinitely;
- Need testing and modelling to forecast behaviour;
- Slot cutting is providing local relief in several dams;
- Requires periodic recutting and reinforcement;
- There may be limits to the number of recuts;
- Long term drying and sealing appears to be the only long term option but opportunities are very limited;
- In worst case, decommissioning or replacement.

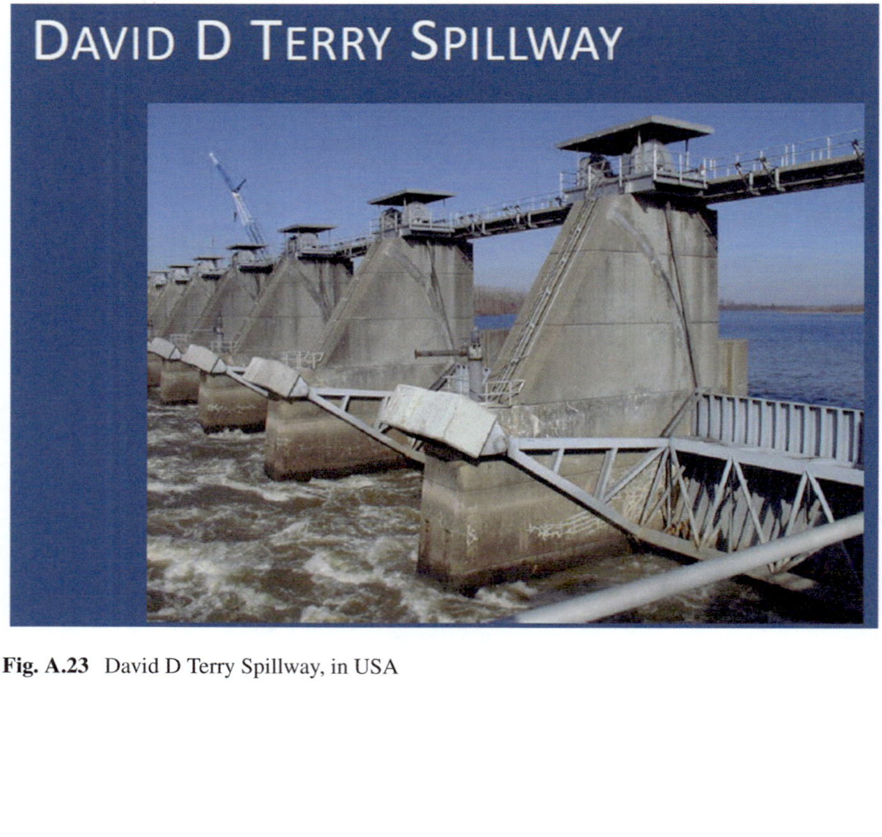

Fig. A.23 David D Terry Spillway, in USA

Fig. A.24 David D Terry Spillway reparation and diagonal cracks

Table A.5 Characteristics of Hiwasse dam and David D Terry Spillway

Official name	Hiwassee Dam
Location	**Cherokee County, North Carolina**, USA
Coordinates	35°9′5″N 84°10′39″W
Construction began	July 15, 1936
Opening date	February 8, 1940
Operator(s)	**Tennessee Valley Authority**
Impounds	**Hiwassee River**
Height	307 ft (94 m)
Length	1376 ft (419 m)
Reservoir	
Creates	**Hiwassee Reservoir**
Total capacity	434000 acre·ft (535000 dam^3)
Catchment area	968 sq mi (2510 km^2)
Power Station	
Installed capacity	185 MW

Fig. A.25 David D Terry Spillway reparation and vertical crack

A.3.6 Prevention of Alkali-Silica Reaction in New Dams

Prevention by to the Components:

Godart et al. [24] referred to a series of specific questions to be considered for the prevention of the appearance of ASR in dams, based on the recommendations of the test methods proposed by RILEM for the characterization of the components [35].

These characteristics analysed in those test methods condition the analysis of the dams in terms of behaviour against ASR. However, there are numerous dams affected by ASR worldwide.

In the Monograph No. 430 of the IETcc-CSIC [33] the following considerations are collected literally, related to dams in Spain.

Peculiarities of the Concrete Components of Dam:

- Typology of aggregates: In many areas of Spain and the Iberian Peninsula in general, there are aggregates with feldspars, such as granites. In these cases, it is necessary to consider, in addition to the reactivity of the aggregates, the potential release of alkalis over time, especially in filtering water conditions such as in dams. On the other hand, many of these aggregates are slow-late reactive, so the expansion tests may be at the limit to be qualified as potentially reactive. Therefore, complementary tests would have to be carried out and the concrete mixtures verified.

- Mineral additions to cement or concrete: Fly ash has been used regularly in dam concrete to obtain suitable characteristics and control the heat of hydration produced, as well as controlling expansion when there are reactive aggregates. However, when this type of addition is in short supply, it is necessary to opt for the use of other types of additions. For the different types of additions, it is necessary to consider their efficiency in controlling expansion and the replacement content required to obtain the same degree of expansion attenuation. In addition, the potential long-term release of alkali from the additions must be considered.
- Peculiarities of the tests for the prevention of expansion in the concrete of dams:
  - *Petrography of the aggregates*: The petrographic analysis of the aggregates may not be very definitive in relation to the ASR, when the standard mentioned in the 2021 Structural Code [18] is used. Therefore, to obtain information on the potential reactivity of alkalis, it is recommended to use the specific petrographic method to analyse these reactions, such as the RILEM AAR-1.1 method [38] or even perform specific analyses on the reactivity index of quartz [32].
  - The design and evaluation of aggregate mixtures with different reactivity can be considered, to obtain non-reactive mixtures. To do this is possible to use the UNE 83968 [50] standard. To test the percentages of suitable quantification of optimum reactive/no reactive aggregate ratio is possible to use the accelerated mortar bar (ASTM C 1260) and/or concrete prism test.
  - Analysis of the efficiency of the additions: The use of additions in the cement or in the concrete moderates the expansion of the concrete with expansive aggregates. However, it is necessary to evaluate the efficiency in moderating the expansion, either of different types of additions or of different percentages of substitution. For this, the test of the UNE 83969 [51] standard can be used.
  - Verification of concrete mixtures: To avoid the expansion of concrete mixtures, it is necessary to verify the concrete mixtures to be used in the dam, for which it is recommended to apply the UNE 83967 [49] test, with an expansion limit of 0.02% at 20 weeks, using potentially leachable supplemental alkalis from the aggregates and additions.

RILEM Recommendations for Concrete Dams and Concrete Hydraulic Structures

The RILEM AAR-7.3 document expands the recommendations of the RILEM AAR-7.1 document, particularizing them for hydraulic structures and for structures exposed to an aggressive environment for a long period of time in general [37].

Dams are considered particularly sensitive structures, so a P4 level of precaution (extraordinary level of precaution) is applied to them, this is due to the great importance of their structural stability, their forecast of a very long life in service and because they are permanent exposed to high moisture content. According to the

P4 level of precaution, two or more of the prevention measures M1, M2, M3 and M4 must be applied.

In addition, some peculiarities of the dams and some specific factors related to these structures and the development of the AAR must be considered:

- Slow reactions can progress and generate significant expansions, due to the long service life and high volume of concrete;
- In large structures, the alkali leaching that can occur in accelerated tests does not occur;
- Dams have a high requirement for structural stability;
- The concrete is permanently in contact with moisture, although there are protection measures that can be applied;
- Dams are often built in remote locations, so there is limited access to concrete components, especially aggregates;
- The horizontal restriction imposed, by the contact of the dam with the rock, in the abutments can cause them to tilt or bend due to expansion, especially in arch dams or arch-gravity dams that have a curved axis;
- The expansions produced in the concrete can cause problems in the connections of the spillways, gates, mechanisms and other installations.

References

1. ASTM C 1260: Standard Test Method for Potential Alkali Reactivity of Aggregates (Mortar-Bar Method). ASTM International, West Conshohocken, USA, 5 pp. (2021)
2. Bérubé, M.A., Frenette, J., Landry, M., Mcpherdan, D., Pedneault, A., Ouellet, S.: Evaluation du potentiel résiduel de réaction et d'expansion du béton en service atteint de réactivité alcalis-silice. Rapport final no. CO-93-04, soumis à Hydro Québec, 151 pp. (1994)
3. Bérubé, M.A., Dorion, J.F., Rivest, M.: Distribution of alkalis in concrete structures affected by alkali-silica reactivity and contribution by Aggregates. In: Proceedings of the 11th International Conference on Alkali-Aggregate reaction (AAR) in concrete, Québec (Canada), pp. 139–149 (2000)
4. Bérubé, M.A., Frenette, J., Pedneault, A., Rivest, M.: Labouratory assessment of the potential rate of ASR expansion of field concrete. Cem. Concr. Aggreg. **24**(1), 13–19 (2002)
5. Bérubé, M.A., Frenette, J., Rivest, M., Vézina, D.: Measurement of the Alkali content of concrete using hot-water extraction. Cem. Concr. Aggreg. **24**(1), 28–36 (2002b)
6. Bund-Länder-Vertreter, et al.: Vertreter des Bundes und der Länder sowie weitere Fachleute. Empfehlungen für die Schadensdiagnose und die Bauliche Erhaltung von AKR-geschädigten Fahrbahndecken aus Beton, Ministerium für Verkehr, Stuttgart, Germany, 51 pp. (2021)
7. Charlwood, R., Bull, S.J.: Workshop: chemical expansion of concrete in dams. In: Hydro 2007, Granada – Spain, ICOLD/SPANCOLD Workshop (2007)
8. Charlwood, R., Sanz, J.M.B.: Chemical expansion of concrete in dams & hydro-electric project. In: ICOLD Symposium, Sofia (2008)
9. Charlwood, R., Buil Sanz, J.M.: Chemical expansion of concrete in dams and hydro-electric projects, workshop report. Hydropower Dams **3** (2008)
10. Charlwood, R.: International Workshop on Managing AAR in Dams, ICOLD 2013 Post Tour & Workshop B4–Southeastern US Dams with Alkali-Aggregate Reactions, August 20, 2013, ICOLD-Annual Meeting 2013 Seattle. WA, USA (2013)
11. Constantiner, D., Diamond, S.: Alkali release from feldspars into pore solutions. Cem. Concr. Res. **33**(4), 549–554 (2003)
12. Bundestag: Antwort auf die kleine Anfrage der Fraktion BÜNDNIS 90/DIE GRÜNEN (Drucksache 16/12024): Zerstörung des Fahrbahnbelages durch die Alkali-Kieselsäure-Reaktion (2009)
13. DafStb: Deutscher Ausschuss für Stahlbeton (Herausgeber): Vorbeugende Maßnahmen gegen schädigende Alkalireaktion im Beton (Alkali-Richtlinie) (2019)
14. Drolet, C., Duchesne, J., Fournier, B.: Effect of alkali release by aggregates on alkali-silica reaction. Constr. Build. Mater. **157**, 263–276 (2017)
15. Drolet, C., Duchesne, J., Fournier, B.: Validation of the alkali contribution by aggregates to the concrete pore solution. Cem. Concr. Res. **98**, 10–23 (2017)

E. Menéndez Méndez and A. Santos Silva (eds.), *Total Alkalis in Concrete—Contribution to Alkali-silica Reaction*, RILEM State-of-the-Art Reports 40, https://doi.org/10.1007/978-3-031-56164-1

16. Dron, R., Privot, F.: Thermodynamic and kinetic approach to the alkali-silica reaction. Part I: Concepts. Cem. Concr. Res. **22**(5), 941–948 (1992)
17. Durand, B.: A note about alkali contribution from aggregates in concrete affected by ASR. In: Proceedings of the 11th International Conference on Alkali-Aggregate Reaction in Concrete, Québec City, Canada, pp. 169–177 (2000)
18. EN 932-3: Tests for general properties of aggregates Part 3: Procedure and terminology for simplified petrographic description, CEN, Brussels, 15 pp. (1996)
19. FGSV: German guideline for building highway pavements since 2018 (2018)
20. Fournier, B., Sanchez, L., Duchesne, J., Goyette, S.: Evaluation of the available alkali content in concrete through a modified hot-water extraction method. In: Proceedings of the 15th International Conference on Alkali-Aggregate Reaction in Concrete, S. Paulo, Brazil, 3–7 July, 10 p. (2016)
21. Fournier, B., Bérubé, M.A., Folliard, K.J., Thomas, M.: Report on the Diagnosis, Prognosis, and Mitigation of Alkali-Silica Reaction (ASR) in Transportation Structures, Federal Highway Administration, USA, 147 pp. (2010)
22. Giebson, C.: Die Alkali-Kieselsäure-Reaktion in Beton für Fahrbahndecken und Flugbetriebsflächen unter Einwirkung alkalihaltiger Enteisungsmittel. Dissertation, Bauhaus-Universität Weimar (http://nbn-resolving.de/urn:nbn:de:gbv:wim2-20131217-20916), 204 pp. (2013)
23. Giebson, C., Voland, K., Ludwig, H.M., Meng, B.: Alkali-silica reaction performance testing of concrete considering external alkalis and preexisting microcracks. Struct. Concr. **18**(2017), 528–538 (2017)
24. Godart, B., de Rooij, M., Wood, J.G.M.: Guide to Diagnosis and Appraisal of AAR Damage to Concrete in Structures: Part 1 Diagnosis (AAR 6.1), 12 RILEM State-of-the-Art Reports, Ed. Springer; ISBN-10:9400796269, ISBN-13:978-9400796263, 89 pp. (2013)
25. ICOLD Bulletin 79: Alkali-Aggregate Reaction in Concrete Dams – Review and Recommendations, Ed. ICOLD (1991)
26. ICOLD: Symposium A, Long-term behaviour of dams. In: International Commission on Large Dams 79th Annual Meeting, Lucerne/Switzerland (2011)
27. ICOLD Bulletin 184: Managing Expansive Chemical Reactions in Concrete Dams, ICOLD, Committee on Concrete Dams (2021)
28. Kermit, P.H.: Determining alkali content in ASR performance-tested concrete. Norwegian University of Science and Technology, 70 pp. (2017)
29. Lagerblad, B.: Alkali release from silicate minerals and alkali-silica reaction in concrete. In: Proceedings of 14th ICAAR, Austin, USA, 6 pp. (2012)
30. Menéndez, E., Prendes, N., Márquez, C., Aldea, B.: Evaluation of granitic aggregates behaviour in relation with the alkaline extraction and compositional change in their phases. In: Proceedings of 14th ICAAR, Austin, USA, 10 pp. (2012)
31. Menéndez, E.: Spanish experience with AAR & ISA, Workshop on Alkali Aggregate Reactions, 20 Aug 2013, 81st Annual Meeting of the International Commission on Large Dams, organized by Robin Charlwood (2013)
32. Menéndez, E., Aldea, B., García-Roves, R.: Prevention, diagnosis and prognosis of alkali-silica reaction, The International Journal of Hydropower and Dams, Shaping the future of hydropower, Seville, Spain, October 2017, Section 8, paper 08.05 (2017)
33. Menéndez, E.: Integral Strategy for the Prevention of the Alkali-Aggregate Reaction (in Spanish), Monografía N° 430 del IETcc-CSIC, ISBN 978-84-00-10546-7 (2019)
34. NB Power, J.F., Hatch, D.C.: Managing AAR at Mactaquac GS, ENBP & Hatch. In: International Workshop on Managing AAR in Dams, ICOLD 2013 Post Tour & Workshop B4–Southeastern US Dams with Alkali-Aggregate Reactions, August 20, 2013, ICOLD-Annual Meeting 2013 Seattle, WA, USA (2013)
35. Nixon, P.J., Sims, I.: RILEM recommendations for prevention of damage by alkli-aggregate reactions in new concrete structures, State-of-the-Art Report of the RILEM Technical Committee-219-ACS, ISBN 978-94-017-7251-8 (2016)
36. Plusquellec, G., Geiker, M.R., Duchesne, J., Lindgård, J., De Weerdt, K.: Review of methods to determine the pH and the free alkali content of the pore solution in concrete. In: Proceedings of 15th ICAAR, S. Paulo, Brazil, 9 pp. (2016)

37. RILEM AAR-7.3: RILEM recommended specification: AAR-7.3 preliminary international specification to minimize damage from alkali-aggregate reactions in concrete - Part 3: concrete dams and other hydro structures. In: Nixon, P., Sims, I. (eds.) RILEM Recommendations for the Prevention of Damage by Alkali-Aggregate Reactions in New Concrete Structures. RILEM State-of-the-Art Reports, vol. 17. Springer, Dordrecht (2016). https://doi.org/10.1007/978-94-017-7252-5_10
38. RILEM AAR-7.1: RILEM recommended specification: AAR-7.1 international specification to minimise damage from alkali reactions in concrete - Part 1: alkali-silica reaction. In: Nixon, P., Sims, I. (eds.) RILEM Recommendations for the Prevention of Damage by Alkali-Aggregate Reactions in New Concrete Structures. RILEM State-of-the-Art Reports, vol. 17. Springer, Dordrecht (2016). https://doi.org/10.1007/978-94-017-7252-5_8
39. Saouma, V.E., Hariri-Ardebili, M.A.: A proposed aging management program for alkali silica reactions in a nuclear power plant. Nucl. Eng. Des. **277**, 248–264 (2014)
40. Seyfarth, K., Giebson, C.: Beurteilung des AKR-Schädigungspotentials von Betonen mittels Klimawechsellagerung. Beton- und Stahlbetonbau (100), Tagungsband Doktorandensymposium 45. Forschungskolloquium des DAfStb., Wien, Österreich, S. 189–192 (2005)
41. Seyfarth, K., Giebson, C., Stark, J.: AKR-performance-Prüfung für Fahrbahndecken aus Beton: Erfahrungen aus Labour und Praxis im Vergleich. 17. Internationale Baustofftagung (ibausil). Weimar, Tagungsbericht **2**, 255–260 (2009)
42. Seyfarth, K., Giebson, C., Stark, J.: Prevention of deleterious ASR by assessing aggregates and specific concrete mixtures. In: Proceedings of the 3rd International Conference on Concrete and Development, Tehran, Iran (2009), pp. 159–169 (2009b)
43. Sims, I., Poole, A.B.: Alkali-Aggregate Reaction in Concrete: A World Re-view. CRC Press/ Balkema, Taylor & Francis Group (2017)
44. Soares, D., Santos Silva, A., Mirão, J., Ramos, V., Fernandes, I., Menéndez, E.: Assessment of alkalis released by aggregates, contribution to the alkalinity increase and AAR development in concrete. In: Second International Dam World Conference, Portugal, Lisbon, LNEC (2015)
45. Soares, D., Santos Silva, A., Mirão, J., Fernandes, I., Menéndez, E.: Study on the factors affecting alkalis release from aggregates into ASR. In: Proceedings of the 15th International Conference on Alkali-Aggregate Reaction in Concrete, S. Paulo, Brazil, 3–7 July, 10 p. (2016)
46. Stark, J., Giebson, C.: Assessing the durability of concrete regarding ASR. In: Malhotra, V.M. (ed.) Proceedings of the 7th CANMET/ACI International Conference on Durability of Concrete, Montreal, Canada, pp. 225–238 (2006)
47. Stark, J., Seyfarth, K.: Performance testing method for durability of concrete using climate simulation. In: Proceedings of the 7th CANMET/ACI International Conference on Durability of Concrete, Montreal, Canada, pp. 305–326 (2006)
48. Thomas, M.D.A., Folliard, K.J., Fournier, B., Rivard, P., Drimalas, T., Garber, S.I.: Methods for Evaluating and Treating ASR-Affected Structures: Results of Field Application and Demonstration Projects Volume II: Details of Field Applications and Analysis, Federal Highway Administration, USA, 320 pp. (2013)
49. UNE 83967: Durabilidad del hormigón. Método de ensayo para evaluación de la expansión potencial por reacción árido-álcali de dosificaciones de hormigón. Método semi-acelerado de prismas de hormigón, UNE, Spain, 10 pp. (2016)
50. UNE 83968: Durabilidad del hormigón. Evaluación de la expansión de probetas de mortero utilizando mezclas de áridos potencialmente reactivos y no reactivos, frente a la reacción álcali-sílice y álcali-silicato. Método acelerado en probetas de mortero, UNE, Spain, 10 pp. (2016)
51. UNE 83969: Durabilidad del hormigón. Evaluación de la expansión de probetas de mortero empleando conglomerantes y áridos potencialmente reactivo frente a la reacción álcali-sílice y álcali-silicato. Método acelerado en probetas de mortero, UNE, Spain, 11 pp. (2017)
52. USCOLD: Dam safety and operations, Ed. USCOLD (1995)

MIX
Papier aus verantwortungsvollen Quellen
Paper from responsible sources
FSC® C105338

If you have any concerns about our products,
you can contact us on
ProductSafety@springernature.com

In case Publisher is established outside the EU,
the EU authorized representative is:
Springer Nature Customer Service Center GmbH
Europaplatz 3, 69115 Heidelberg, Germany

Printed by Libri Plureos GmbH
in Hamburg, Germany